T0133171

Boundary Methods

Elements, Contours, and Nodes

MECHANICAL ENGINEERING
A Series of Textbooks and Reference Books

Founding Editor

L. L. Faulkner

*Columbus Division, Battelle Memorial Institute
and Department of Mechanical Engineering
The Ohio State University
Columbus, Ohio*

Boundary Methods

Elements, Contours, and Nodes

Subrata Mukherjee
Cornell University
Ithaca, New York, U.S.A.

Yu Xie Mukherjee
Cornell University
Ithaca, New York, U.S.A.

Taylor & Francis
Taylor & Francis Group

Boca Raton London New York Singapore

A CRC title, part of the Taylor & Francis imprint, a member of the
Taylor & Francis Group, the academic division of T&F Informa plc.

Published in 2005 by
CRC Press
Taylor & Francis Group
6000 Broken Sound Parkway NW
Boca Raton, FL 33487-2742

International Standard Book Number-10: 0-8247-2599-9 (Hardcover)
International Standard Book Number-13: 978-0-8247-2599-0 (Hardcover)
Library of Congress Card Number 2004063489

This book contains information obtained from authentic and highly regarded sources. Reprinted material is quoted with permission, and sources are indicated. A wide variety of references are listed. Reasonable efforts have been made to publish reliable data and information, but the author and the publisher cannot assume responsibility for the validity of all materials or for the consequences of their use.

Library of Congress Cataloging-in-Publication Data

Mukherjee, Subrata.
    Boundary methods : elements, contours, and nodes / Subrata Mukherjee and Yu Mukherjee.
        p. cm. -- (Mechanical engineering ; 185)
    ISBN 0-8247-2599-9 (alk. paper)
    1. Boundary element methods. I. Mukherjee, Yu. II. Title. III. Mechanical engineering (Marcel Dekker, Inc.) ; 185.

TA347.B69M83 2005
621'.01'51535--dc22   2004063489

Taylor & Francis Group
is the Academic Division of T&F Informa plc.

Visit the Taylor & Francis Web site at
http://www.taylorandfrancis.com

and the CRC Press Web site at
http://www.crcpress.com

To our boys
Anondo and Alok

and

To Yu's teacher, Professor Zhicheng Xie of Tsinghua University,
a distinguished scholar
who has dedicated himself to China.

PREFACE

The general subject area of concern to this book is computational science and engineering, with applications in potential theory and in solid mechanics (linear elasticity). This field has undergone a revolution during the past several decades along with the exponential growth of computational power and memory. Problems that were too large for main frame computers 15 or 20 years ago can now be routinely solved on desktop personal computers.

There are several popular computational methods for solving problems in potential theory and linear elasticity. The most popular, versatile and most commonly used is the finite element method (FEM). Many hundreds of books already exist on the subject and new books get published frequently on a regular basis. Another popular method is the boundary element method (BEM). Compared to the FEM, we view the BEM as a niche method, in that it is particularly well suited, from the point of view of accuracy as well as computational efficiency, for linear problems. The principal advantage of the BEM, relative to the FEM, is its dimensionality advantage. The FEM is a domain method that requires discretization of the entire domain of a body while the BEM, for linear problems, only requires discretization of its bounding surface.

The process of discretization (or meshing) of a three-dimensional (3-D) object of complex shape is a popular research area in computational geometry. Even though great strides have been made in recent years, meshing, for many applications, still remains an arduous task. During the past decade, mesh-free (also called mesh-less) methods have become a popular research area in computational mechanics. The main purpose here is to substantially simplify the task of meshing of an object. Advantages of mesh-free methods become more pronounced, for example, for problems involving optimal shape design or adaptive meshing, since many remeshings must be typically carried out for such problems. One primary focus of this book is a marriage of these two ideas, i.e. a discussion of a boundary-based mesh-free method - the boundary node method (BNM) - which combines the dimensionality advantage of the BEM with the ease of discretization of mesh-free methods.

Following an introductory chapter, this book consists of three parts related to the boundary element, boundary contour and boundary node methods. The first part is short, in order not to duplicate information on the BEM that is already available in many books on the subject. Only some novel topics related to the BEM are presented here. The second part is concerned with the boundary contour method (BCM). This method is a novel variant of the BEM in that it further reduces the dimensionality of a problem. Only one-dimensional line integrals need to be numerically computed when solving three-dimensional problems in linear elasticity by the BCM. The third part is concerned with the boundary node method (BNM). The BNM combines the BEM with moving least-squares (MLS) approximants, thus producing a mesh-free boundary-only method. In addition to the solution of 3-D problems, Part II of the book on the BCM presents shape sensitivity analysis, shape optimization, and error estimation and adaptivity; while Part III on the BNM includes error analysis and adaptivity.

This book is written in the style of a research monograph. Each topic is clearly introduced and developed. Numerical results for selected problems appear throughout the book, as do references to related work (research publications and books).

This book should be of great interest to graduate students, researchers and practicing engineers in the field of computational mechanics; and to others interested in the general areas of computational mathematics, science and engineering. It should also be of value to advanced undergraduate students who are interested in this field.

We wish to thank a number of people and organizations who have contributed in various ways to making this book possible. Two of Subrata's former graduate students, Glaucio Paulino and Mandar Chati, as well as Yu's associate Xiaolan Shi, have made very significant contributions to the research that led to this book. Sincere thanks are expressed to Subrata's former graduate students Govind Menon and Ramesh Gowrishankar, to one of his present graduate students, Srinivas Telukunta, and to Vasanth Kothnur, for their contributions to the BNM. Earlin Lutz, Anantharaman Nagarajan and Anh-Vu Phan have significantly contributed to the early development of the BCM; while Subrata's just-graduated student Zhongping Bao has made excellent contributions to the research on micro-electro-mechanical systems (MEMS) by the BEM. Sincere thanks are expressed to our dear friend Ashim Datta for his help and encouragement throughout the writing of this book.

Much of the research presented here has been financially supported by the National Science Foundation and Ford Motor Company, and this support is gratefully acknowledged. Most of the figures and tables in this book have been published before in journals. They were all originally created by the authors of this book, together with their coauthors. These items have been printed here by permission of the original copyright owner (i.e. the publishers of the appropriate journal), and this permission is very much appreciated. The original source has been acknowledged in this book at the end of the caption for each item.

Subrata and Yu Mukherjee
Ithaca, New York
October 2004

Contents

INTRODUCTION TO BOUNDARY METHODS

This chapter provides a brief introduction to various topics that are of interest in this book.

Boundary Element Method

Boundary integral equations (BIE), and the boundary element method (BEM), based on BIEs, are mature methods for numerical analysis of a large variety of problems in science and engineering. The standard BEM for linear problems has the well-known dimensionality advantage in that only the two-dimensional (2-D) bounding surface of a three-dimensional (3-D) body needs to be meshed when this method is used. Examples of books on the subject, published during the last 15 years, are Banerjee [4], Becker [9], Bonnet [14], Brebbia and Dominguez [16], Chandra and Mukherjee [22], Gaul et al. [47], Hartmann [62], Kane [68] and París and Cañas [121]. BEM topics of interest in this book are finite parts (FP) in Chapter 1, error estimation in Chapter 2 and thin features (cracks and thin objects) in Chapter 3.

Hypersingular Boundary Integral Equations

Hypersingular boundary integral equations (HBIEs) are derived from a differentiated version of the usual boundary integral equations (BIEs). HBIEs have diverse important applications and are the subject of considerable current research (see, for example, Krishnasamy et al. [76], Tanaka et al. [162], Paulino [122] and Chen and Hong [30] for recent surveys of the field). HBIEs, for example, have been employed for the evaluation of boundary stresses (e.g. Guiggiani [60], Wilde and Aliabadi [173], Zhao and Lan [185], Chati and Mukherjee [24]), in wave scattering (e.g. Krishnasamy et al. [75]), in fracture mechanics (e.g. Cruse [38], Gray et al. [54], Lutz et al. [89], Paulino [122], Gray and Paulino [58], Mukherjee et al. [110]), to obtain symmetric Galerkin boundary element formulations (e.g. Bonnet [14], Gray et al. [55], Gray and Paulino ([56], [57]), to evaluate nearly singular integrals (Mukherjee et al. [104]), to obtain the hypersingular boundary contour method (Phan et al. [131], Mukherjee and Mukherjee [99]), to obtain the hypersingular boundary node method (Chati et al. [27]), and for error analysis (Paulino et al. [123], Menon [95], Menon et al. [96], Chati et al. [27], Paulino and Gray [125]) and adaptivity [28].

An elegant approach of regularizing singular and hypersingular integrals, using simple solutions, was first proposed by Rudolphi [143]. Several researchers have used this idea to regularize hypersingular integrals before collocating an HBIE at a regular boundary point. Examples are Cruse and Richardson [39], Lutz et al. [89], Poon et al. [138], Mukherjee et al. [110] and Mukherjee [106]. The relationship between finite parts of strongly singular and hypersingular integrals, and the HBIE, is discussed in [168], [101] and [102]. A lively debate (e.g. [92], [39]), on smoothness requirements on boundary variables for collocating an HBIE on the boundary of a body, has apparently been concluded recently [93]. An alternative way of satisfying this smoothness requirement is the use of the hypersingular boundary node method (HBNM).

Mesh-Free Methods

Mesh-free (also called meshless) methods [82], that only require points rather than elements to be specified in the physical domain, have tremendous potential advantages over methods such as the finite element method (FEM) that require discretization of a body into elements.

The idea of moving least squares (MLS) interpolants, for curve and surface fitting, is described in a book by Lancaster and Salkauskas [78]. Nayroles et al. [117] proposed a coupling of MLS interpolants with Galerkin procedures in order to solve boundary value problems. They called their method the diffuse element method (DEM) and applied it to two-dimensional (2-D) problems in potential theory and linear elasticity.

During the relatively short span of less than a decade, great progress has been made in solid mechanics applications of mesh-free methods. Mesh-free methods proposed to date include the element-free Galerkin (EFG) method [10, 11, 12, 13, 67, 174, 175, 176, 108], the reproducing-kernel particle method (RKPM) [83, 84], $h - p$ clouds [42, 43, 120], the meshless local Petrov-Galerkin (MLPG) approach [3], the local boundary integral equation (LBIE) method [152, 188], the meshless regular local boundary integral equation (MRLBIE) method [189], the natural element method (NEM) [158, 160], the generalized finite element method (GFEM) [157], the extended finite element method (X-FEM) [97, 41, 159], the method of finite spheres (MFS) [40], the finite cloud method (FCM) [2], the boundary cloud method (BCLM) [79, 80], the boundary point interpolation method (BPIM) [82], the boundary-only radial basis function method (BRBFM) [32] and the boundary node method (BNM) [107, 72, 25, 26, 27, 28, 52].

Boundary Node Method

S. Mukherjee, together with his research collaborators, has recently pioneered a new computational approach called the boundary node method (BNM) [26, 25, 27, 28, 72, 107]. Other examples of boundary-based meshless methods are the boundary cloud method (BCLM) [79, 80], the boundary point interpolation method (BPIM) [82], the boundary only radial basis function method (BRBFM) [32] and the local BIE (LBIE) [188] approach. The LBIE, however, is not a boundary method since it requires evaluation of integrals over certain surfaces (called L_s in [188]) that can be regarded as "closure surfaces" of boundary elements.

The BNM is a combination of the MLS interpolation scheme and the standard boundary integral equation (BIE) method. The method divorces the traditional coupling between spatial discretization (meshing) and interpolation (as commonly practiced in the FEM or in the BEM). Instead, a "diffuse" interpolation, based on MLS interpolants, is used to represent the unknown functions; and surface cells, with a very flexible structure (e.g. any cell can be arbitrarily subdivided without affecting its neighbors [27]) are used for integration. Thus, the BNM retains the *meshless attribute of the EFG method and the dimensionality advantage of the BEM*. As a consequence, the BNM only requires the specification of *points on the 2-D bounding surface* of a 3-D body (including crack faces in fracture mechanics problems), together with surface cells for integration, thereby practically eliminating the meshing problem (see Figures i and ii). The required cell structure is analogous to (but not the same as) a tiling [139]. The only requirements are that the intersection of any two surface cells is the null set and that the union of all the cells is the bounding surface of the body. In contrast, the FEM needs volume meshing, the BEM needs surface meshing, and the EFG needs points throughout the domain of a body.

It is important to point out another important advantage of MLS interpolants. They can be easily designed to be sufficiently smooth to suit a given purpose, e.g. they can be made C^1 or higher [10] in order to collocate the HBNM at a point on the boundary of a body.

The BNM is described in Chapters 8 and 9 of this book.

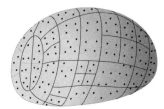   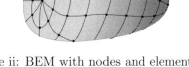

Figure i: BNM with nodes and cells Figure ii: BEM with nodes and elements
        (from [28])   (from [28])

Boundary Contour Method

The Method

The usual boundary element method (BEM), for three-dimensional (3-D) linear elasticity, requires numerical evaluations of surface integrals on boundary elements on the surface of a body (see, for example, [98]). [115] (for 2-D linear elasticity) and [116] (for 3-D linear elasticity) have recently proposed a novel approach, called the boundary contour method (BCM), that achieves a further reduction in dimension! The BCM, for 3-D linear elasticity problems, only requires numerical evaluation of line integrals over the closed bounding contours of the usual (surface) boundary elements.

The central idea of the BCM is the exploitation of the divergence-free property of the usual BEM integrand and a very useful application of Stokes' theorem, to analytically convert surface integrals on boundary elements to line integrals on closed contours that bound these elements. [88] first proposed an application of this idea for the Laplace equation and Nagarajan et al. generalized this idea to linear elasticity. Numerical results for two-dimensional (2-D) problems, with linear boundary elements, are presented in [115], while results with quadratic boundary elements appear in [129]. Three-dimensional elasticity problems, with quadratic boundary elements, are the subject of [116] and [109]. Hypersingular boundary contour formulations, for two-dimensional [131] and three-dimensional [99] linear elasticity, have been proposed recently. A symmetric Galerkin BCM for 2-D linear elasticity appears in [119]. Recent work on the BCM is available in [31, 134, 135, 136, 186].

The BCM is described in Chapter 4 of this book.

Shape Sensitivity Analysis with the BCM and the HBCM

Design sensitivity coefficients (DSCs), which are defined as rates of change of physical response quantities with respect to changes in design variables, are useful for various applications such as in judging the robustness of a given design, in reliability analysis and in solving inverse and design optimization problems. There are three methods for design sensitivity analysis (e.g. [63]), namely, the finite difference approach (FDA), the adjoint structure approach (ASA) and the direct differentiation approach (DDA). The DDA is of interest in this work.

The goal of obtaining BCM sensitivity equations can be achieved in two equivalent ways. In the 2-D work by [130], design sensitivities are obtained by first converting the discretized BIEs into their boundary contour version, and then applying the DDA (using the concept of the material derivative) to this BCM version. This approach, while relatively straightforward in principle, becomes extremely algebraically intensive for 3-D elasticity problems. [100] offers a novel alternative derivation, using the opposite process, in which the DDA is first applied to the regularized BIE and then the resulting equations are converted to their boundary contour version. It is important to point out that this process of converting the sensitivity BIE into a BCM form is quite challenging. This new derivation, for sensitivities of surface variables [100], as well as for internal variables [103], for 3-D elasticity problems, is presented in Chapter 5 of this book. The reader is referred to [133] for a corresponding derivation for 2-D elasticity

Shape Optimization with the BCM

Shape optimization refers to the optimal design of the shape of structural components and is of great importance in mechanical engineering design. A typical gradient-based shape optimization procedure is an iterative process in which iterative improvements are carried out over successive designs until an optimal design is accepted. A domain-based method such as the finite element method (FEM) typically requires discretization of the entire domain of a body many times during this iterative process. The BEM, however, only requires surface discretization, so that mesh generation and remeshing procedures can be carried out much more easily for the BEM than for the FEM. Also, surface stresses are typically obtained very accurately in the BEM. As a result, the BEM has been a popular method for shape optimization in linear mechanics. Some examples are references [33], [145], [178], [144], [169], [177], [161] and the book [184].

In addition to having the same meshing advantages as the usual BEM, the BCM, as explained above, offers a further reduction in dimension. Also, surface stresses can be obtained very easily and accurately by the BCM without the need for additional shape function differentiation as is commonly required with the BEM. These properties make the BCM very attractive as the computational engine for stress analysis for use in shape optimization. Shape optimization in 2-D linear elasticity, with the BCM, has been presented by [132]. The corresponding 3-D problem is presented in [150] and is discussed in Chapter 6.

Error Estimation and Adaptivity

A particular strength of the finite element method (FEM) is the well-developed theory of error estimation, and its use in adaptive methods (see, for example, Ciarlet [34], Eriksson et al. [44]). In contrast, error estimation in the boundary element method (BEM) is a subject that has attracted attention mainly over the past decade, and much work remains to be done. For recent surveys on error estimation and adaptivity in the BEM, see Sloan [155], Kita and Kamiya [70], Liapis [81] and Paulino et al. [124].

Many error estimators in the BEM are essentially heuristic and, unlike for the FEM, theoretical work in this field has been quite limited. Rank [140] proposed error indicators and an adaptive algorithm for the BEM using techniques similar to those used in the FEM. Most notable is the work of Yu and Wendland [171, 172, 181, 182], who have presented local error estimates based on a linear error-residual relation that is very effective in the FEM. More recently, Carstensen et al. [18, 21, 19, 20] have presented error estimates for the BEM analogous to the approach of Eriksson [44] for the FEM. There are numerous stumbling blocks in the development of a satisfactory theoretical analysis of a generic boundary value problem (BVP). First, theoretical analyses are easiest for Galerkin schemes, but most engineering codes, to date, use collocation-based methods (see, for example, Banerjee [4]). Though one can view collocation schemes as variants of Petrov-Galerkin methods, and, in fact, numerous theoretical analyses exist for collocation methods (see, for example, references in [155]), the mathematical analysis for this class of problems is difficult. Theoretical analyses for mixed boundary conditions are limited and involved (Wendland et al. [170]) and the presence of corners and cracks has been a source of challenging problems for many years (Sloan [155], Costabel and Stephan [35], Costabel et al. [36]). Of course, problems with corners and mixed boundary conditions are the ones of most practical interest, and for such situations one has to rely mostly on numerical experiments.

During the past few years, there has been a marked interest, among mathematicians in the field, in extending analyses for the BEM with singular integrals to hypersingular integrals ([21, 19, 156, 45]. For instance, Feistauer et al. [45] have studied the solution of the exterior Neumann problem for the Helmholtz equation formulated as an HBIE. Their paper contains a rigorous analysis of hypersingular integral equations and addresses the problem of noncompatibility of the residual norm, where additional hypotheses are needed to design a practical error estimate. These authors use residuals to estimate the error, but they do not use the BIE and the HBIE simultaneously. Finally, Goldberg and Bowman [51] have used superconvergence of the Sloan iterate [153, 154] to show the asymptotic equivalence of the error and the residual. They have used Galerkin methods, an iteration scheme that uses the same integral equation for the approximation and for the iterates, and usual residuals in their work.

Paulino [122] and Paulino et al. [123] first proposed the idea of obtaining a hypersingular residual by substituting the BEM solution of a problem into the hypersingular BEM (HBEM) for the same problem; and then using this residual as an element error estimator in the BEM. It has been proved that ([95], [96], [127]), under certain conditions, this residual is related to a measure of the local error on a boundary element, and has been used to postulate local error estimates on that element. This idea has been applied to the collocation BEM ([123], [96], [127]) and to the symmetric Galerkin BEM ([125]). Recently, residuals have been obtained in the context of the BNM [28] and used to obtain local error estimates (at the element level) and then to drive an h-adaptive mesh refinement process. An analogous approach for error estimation and h-adaptivity, in the context of the BCM, is described in [111]. Ref. [91] has a bibliography of work on mesh generation and refinement up to 1993.

Error analysis with the BEM is presented in Chapter 2, while error analysis and adaptivity in the context of the BCM and the BNM are discussed in Chapter 7, and Chapters 10, 11, respectively, of this book.

Part I

SELECTED TOPICS IN BOUNDARY ELEMENT METHODS

Chapter 1

BOUNDARY INTEGRAL EQUATIONS

Integral equations, usual as well as hypersingular, for internal and boundary points, for potential theory in three dimensions, are first presented in this chapter. This is followed by their linear elasticity counterparts. The evaluation of finite parts (FPs) of some of these equations, when the source point is an irregular boundary point (situated at a corner on a one-dimensional plane curve or at a corner or edge on a two-dimensional surface), is described next.

1.1 Potential Theory in Three Dimensions

The starting point is Laplace's equation in three dimensions (3-D) governing a potential function $u(x_1, x_2, x_3) \in B$, where B is a bounded region (also called the body):

$$\nabla^2 u(x_1, x_2, x_3) \equiv \frac{\partial^2 u}{\partial x_1^2} + \frac{\partial^2 u}{\partial x_2^2} + \frac{\partial^2 u}{\partial x_3^2} = 0 \tag{1.1}$$

along with prescribed boundary conditions on the bounding surface ∂B of B.

1.1.1 Singular Integral Equations

Referring to Figure 1.1, let $\boldsymbol{\xi}$ and $\boldsymbol{\eta}$ be (internal) source and field points $\in B$ and $\mathbf{x}$ and $\mathbf{y}$ be (boundary) source and field points $\in \partial B$, respectively. (Source and field points are also referred to as p and q (for internal points) and as P and Q (for boundary points), respectively, in this book).

The well-known integral representation for (1.1), at an internal point $\boldsymbol{\xi} \in B$, is:

$$u(\boldsymbol{\xi}) = \int_{\partial B} [G(\boldsymbol{\xi}, \mathbf{y})\tau(\mathbf{y}) - F(\boldsymbol{\xi}, \mathbf{y})u(\mathbf{y})]dS(\mathbf{y}) \tag{1.2}$$

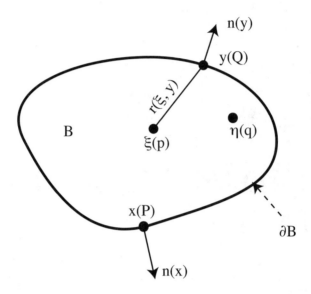

Figure 1.1: Notation used in integral equations (from [6])

An infinitesimal surface area on ∂B is $d\mathbf{S} = dS\mathbf{n}$, where $\mathbf{n}$ is the unit outward normal to ∂B at a point on it and $\tau = \partial u/\partial n$. The kernels are written in terms of source and field points $\boldsymbol{\xi} \in B$ and $\mathbf{y} \in \partial B$. These are :

$$G(\boldsymbol{\xi}, \mathbf{y}) = \frac{1}{4\pi r(\boldsymbol{\xi}, \mathbf{y})} \tag{1.3}$$

$$F(\boldsymbol{\xi}, \mathbf{y}) = \frac{\partial G(\boldsymbol{\xi}, \mathbf{y})}{\partial n(\mathbf{y})} = \frac{(\xi_i - y_i)n_i(\mathbf{y})}{4\pi r^3(\boldsymbol{\xi}, \mathbf{y})} \tag{1.4}$$

in terms of $r(\boldsymbol{\xi}, \mathbf{y})$, the Euclidean distance between the source and field points $\boldsymbol{\xi}$ and $\mathbf{y}$. Unless specified otherwise, the range of indices in these and all other equations in this chapter is 1,2,3.

An alternative form of equation (1.2) is:

$$u(\boldsymbol{\xi}) = \int_{\partial B} [G(\boldsymbol{\xi}, \mathbf{y})u_{,k}(\mathbf{y}) - H_k(\boldsymbol{\xi}, \mathbf{y})u(\mathbf{y})]\mathbf{e}_k \cdot d\mathbf{S}(\mathbf{y}) \tag{1.5}$$

where $\mathbf{e}_k$, $k = 1, 2, 3$, are the usual Cartesian unit vectors, $\mathbf{e}_k \cdot d\mathbf{S}(\mathbf{y}) = n_k(\mathbf{y})dS(\mathbf{y})$, and:

$$H_k(\boldsymbol{\xi}, \mathbf{y}) = \frac{(\xi_k - y_k)}{4\pi r^3(\boldsymbol{\xi}, \mathbf{y})} \tag{1.6}$$

The boundary integral equation (BIE) corresponding to (1.2) is obtained by taking the limit $\boldsymbol{\xi} \to \mathbf{x}$. A regularized form of the resulting equation is:

$$0 = \int_{\partial B} [G(\mathbf{x}, \mathbf{y})\tau(\mathbf{y}) - F(\mathbf{x}, \mathbf{y})\{u(\mathbf{y}) - u(\mathbf{x})\}]dS(\mathbf{y}) \tag{1.7}$$

with an alternate form (from (1.5)):

$$0 = \int_{\partial B} [G(\mathbf{x}, \mathbf{y}) u_{,k}(\mathbf{y}) - H_k(\mathbf{x}, \mathbf{y})\{u(\mathbf{y}) - u(\mathbf{x})\}] e_k \cdot d\mathbf{S}(\mathbf{y}) \qquad (1.8)$$

1.1.2 Hypersingular Integral Equations

Equation (1.2) can be differentiated at an internal source point $\boldsymbol{\xi}$ to obtain the gradient $\frac{\partial u}{\partial \xi_m}$ of the potential u. The result is:

$$\frac{\partial u(\boldsymbol{\xi})}{\partial \xi_m} = \int_{\partial B} \left[\frac{\partial G(\boldsymbol{\xi}, \mathbf{y})}{\partial \xi_m} \tau(\mathbf{y}) - \frac{\partial F(\boldsymbol{\xi}, \mathbf{y})}{\partial \xi_m} u(\mathbf{y}) \right] dS(\mathbf{y}) \qquad (1.9)$$

An interesting situation arises when one takes the limit $\boldsymbol{\xi} \to \mathbf{x}$ ($\mathbf{x}$ can even be an irregular point on ∂B but one must have $u(\mathbf{y}) \in C^{1,\alpha}$ at $\mathbf{y} = \mathbf{x}$) in equation (1.9). As discussed in detail in Section 1.4.2, one obtains:

$$\frac{\partial u(\mathbf{x})}{\partial x_m} = \fint_{\partial B} \left[\frac{\partial G(\mathbf{x}, \mathbf{y})}{\partial x_m} \tau(\mathbf{y}) - \frac{\partial F(\mathbf{x}, \mathbf{y})}{\partial x_m} u(\mathbf{y}) \right] dS(\mathbf{y}) \qquad (1.10)$$

where the symbol $\fint$ denotes the finite part (FP) of the integral. Equation (1.10) is best regularized before computations are carried out. The regularized version given below is applicable even at an irregular boundary point $\mathbf{x}$ provided that $u(\mathbf{y}) \in C^{1,\alpha}$ at $\mathbf{y} = \mathbf{x}$. This is:

$$
\begin{aligned}
0 \;=\; & \int_{\partial B} \frac{\partial G(\mathbf{x}, \mathbf{y})}{\partial x_m} \left[u_{,p}(\mathbf{y}) - u_{,p}(\mathbf{x}) \right] n_p(\mathbf{y}) dS(\mathbf{y}) \\
& - \int_{\partial B} \frac{\partial F(\mathbf{x}, \mathbf{y})}{\partial x_m} \left[u(\mathbf{y}) - u(\mathbf{x}) - u_{,p}(\mathbf{x})(y_p - x_p) \right] dS(\mathbf{y}) \qquad (1.11)
\end{aligned}
$$

An alternative form of (1.11), valid at a regular boundary point $\mathbf{x}$, [76] is:

$$
\begin{aligned}
0 \;=\; & \int_{\partial B} \frac{\partial G(\mathbf{x}, \mathbf{y})}{\partial x_m} \left[\tau(\mathbf{y}) - \tau(\mathbf{x}) \right] dS(\mathbf{y}) \\
& - u_{,k}(\mathbf{x}) \int_B \frac{\partial G(\mathbf{x}, \mathbf{y})}{\partial x_m} \left[n_k(\mathbf{y}) - n_k(\mathbf{x}) \right] dS(\mathbf{y}) \\
& - \int_{\partial B} \frac{\partial F(\mathbf{x}, \mathbf{y})}{\partial x_m} \left[u(\mathbf{y}) - u(\mathbf{x}) - u_{,p}(\mathbf{x})(y_p - x_p) \right] dS(\mathbf{y}) \qquad (1.12)
\end{aligned}
$$

Carrying out the inner product of (1.12) with the source point normal $\mathbf{n}(\mathbf{x})$, one gets:

$$0 \;=\; \int_{\partial B} \frac{\partial G(\mathbf{x}, \mathbf{y})}{\partial n(\mathbf{x})} \left[\tau(\mathbf{y}) - \tau(\mathbf{x}) \right] dS(\mathbf{y})$$

$$- \quad u_{,k}(\mathbf{x}) \int_B \frac{\partial G(\mathbf{x}, \mathbf{y})}{\partial n(\mathbf{x})} \Big[n_k(\mathbf{y}) - n_k(\mathbf{x}) \Big] dS(\mathbf{y})$$

$$- \quad \int_{\partial B} \frac{\partial F(\mathbf{x}, \mathbf{y})}{\partial n(\mathbf{x})} \Big[u(\mathbf{y}) - u(\mathbf{x}) - u_{,p}(\mathbf{x})(y_p - x_p) \Big] dS(\mathbf{y}) \qquad (1.13)$$

1.1.2.1   Potential gradient on the bounding surface

The gradient of the potential function is required in the regularized HBIEs (1.11 - 1.13). For potential problems, the gradient (at a regular boundary point) can be written as,

$$\nabla u = \tau \mathbf{n} + \frac{\partial u}{\partial s_1} \mathbf{t}_1 + \frac{\partial u}{\partial s_2} \mathbf{t}_2 \qquad (1.14)$$

where $\tau = \partial u / \partial n$ is the flux, $\mathbf{n}$ is the unit normal, $\mathbf{t}_1, \mathbf{t}_2$ are the appropriately chosen unit vectors in two orthogonal tangential directions on the surface of the body, and $\partial u / \partial s_i, i = 1, 2$ are the tangential derivatives of u (along $\mathbf{t}_1$ and $\mathbf{t}_2$) on the surface of the body.

1.2   Linear Elasticity in Three Dimensions

The starting point is the Navier-Cauchy equation governing the displacement $\mathbf{u}(x_1, x_2, x_3)$ in a homogeneous, isotropic, linear elastic solid occupying the bounded 3-D region B with boundary ∂B; in the absence of body forces:

$$0 = u_{i,jj} + \frac{1}{1 - 2\nu} u_{k,ki} \qquad (1.15)$$

along with prescribed boundary conditions that involve the displacement and the traction $\boldsymbol{\tau}$ on ∂B. The components τ_i of the traction vector are:

$$\tau_i = \lambda u_{k,k} n_i + \mu(u_{i,j} + u_{j,i}) n_j \qquad (1.16)$$

In equations (1.15) and (1.16), ν is Poisson's ratio and λ and μ are Lamé constants. As is well known, μ is the shear modulus of the material and is also called G in this book. Finally, the Young's modulus is denoted as E.

1.2.1   Singular Integral Equations

The well-known integral representation for (1.15), at an internal point $\boldsymbol{\xi} \in B$ (Rizzo [141]) is:

$$u_k(\boldsymbol{\xi}) = \int_{\partial B} [U_{ik}(\boldsymbol{\xi}, \mathbf{y}) \tau_i(\mathbf{y}) - T_{ik}(\boldsymbol{\xi}, \mathbf{y}) u_i(\mathbf{y})] \, dS(\mathbf{y}) \qquad (1.17)$$

where u_k and τ_k are the components of the displacement and traction respectively, and the well-known Kelvin kernels are:

$$U_{ik} = \frac{1}{16\pi(1-\nu)Gr}[(3-4\nu)\delta_{ik} + r_{,i}r_{,k}] \tag{1.18}$$

$$T_{ik} = -\frac{1}{8\pi(1-\nu)r^2}\left[\{(1-2\nu)\delta_{ik} + 3r_{,i}r_{,k}\}\frac{\partial r}{\partial n} + (1-2\nu)(r_{,i}n_k - r_{,k}n_i)\right] \tag{1.19}$$

In the above, δ_{ik} denotes the Kronecker delta and, as before, the normal $\mathbf{n}$ is defined at the (boundary) field point $\mathbf{y}$. A comma denotes a derivative with respect to a field point, i.e.

$$r_{,i} = \frac{\partial r}{\partial y_i} = \frac{y_i - \xi_i}{r} \tag{1.20}$$

An alternative form of equation (1.17) is:

$$u_k(\boldsymbol{\xi}) = \int_{\partial B}[U_{ik}(\boldsymbol{\xi},\mathbf{y})\sigma_{ij}(\mathbf{y}) - \Sigma_{ijk}(\boldsymbol{\xi},\mathbf{y})u_i(\mathbf{y})]\,\mathbf{e}_j \cdot d\mathbf{S}(\mathbf{y}) \tag{1.21}$$

where $\boldsymbol{\sigma}$ is the stress tensor, $\tau_i = \sigma_{ij}n_j$ and $T_{ik} = \Sigma_{ijk}n_j$. (Please note that $\mathbf{e}_j \cdot d\mathbf{S}(\mathbf{y}) = n_j(\mathbf{y})dS(\mathbf{y})$). The explicit form of the kernel $\boldsymbol{\Sigma}$ is:

$$\begin{aligned}\Sigma_{ijk} &= E_{ijmn}\frac{\partial U_{km}}{\partial y_n} \\ &= -\frac{1}{8\pi(1-\nu)r^2}[\,(1-2\nu)(r_{,i}\delta_{jk} + r_{,j}\delta_{ik} - r_{,k}\delta_{ij}) + 3r_{,i}r_{,j}r_{,k}\,]\end{aligned} \tag{1.22}$$

where $\mathbf{E}$ is the elasticity tensor (for isotropic elasticity):

$$E_{ijmn} = \lambda\delta_{ij}\delta_{mn} + \mu[\delta_{im}\delta_{jn} + \delta_{in}\delta_{jm}] \tag{1.23}$$

The boundary integral equation (BIE) corresponding to (1.17) is obtained by taking the limit $\boldsymbol{\xi} \to \mathbf{x}$. The result is:

$$\begin{aligned}u_k(\mathbf{x}) &= \lim_{\boldsymbol{\xi}\to\mathbf{x}}\int_{\partial B}[U_{ik}(\boldsymbol{\xi},\mathbf{y})\tau_i(\mathbf{y}) - T_{ik}(\boldsymbol{\xi},\mathbf{y})u_i(\mathbf{y})]\,dS(\mathbf{y}) \\ &= \fint_{\partial B}[U_{ik}(\mathbf{x},\mathbf{y})\tau_i(\mathbf{y}) - T_{ik}(\mathbf{x},\mathbf{y})u_i(\mathbf{y})]\,dS(\mathbf{y})\end{aligned} \tag{1.24}$$

where the symbol $\fint_{\partial B}$ denotes the finite part of the appropriate integral (see Section 1.4).

A regularized form of equation (1.24) is:

$$0 = \int_{\partial B}[U_{ik}(\mathbf{x},\mathbf{y})\tau_i(\mathbf{y}) - T_{ik}(\mathbf{x},\mathbf{y})\{u_i(\mathbf{y}) - u_i(\mathbf{x})\}]dS(\mathbf{y}) \tag{1.25}$$

with an alternate form (from (1.21)):

$$0 = \int_{\partial B} [U_{ik}(\mathbf{x}, \mathbf{y})\sigma_{ij}(\mathbf{y}) - \Sigma_{ijk}(\mathbf{x}, \mathbf{y})\{u_i(\mathbf{y}) - u_i(\mathbf{x})\}]e_j \cdot d\mathbf{S}(\mathbf{y}) \qquad (1.26)$$

1.2.2 Hypersingular Integral Equations

Equation (1.17) can be differentiated at an internal source point $\boldsymbol{\xi}$ to obtain the displacement gradient at this point:

$$\frac{\partial u_k(\boldsymbol{\xi})}{\partial \xi_m} = \int_{\partial B} \left[\frac{\partial U_{ik}}{\partial \xi_m}(\boldsymbol{\xi}, \mathbf{y})\tau_i(\mathbf{y}) - \frac{\partial T_{ik}}{\partial \xi_m}(\boldsymbol{\xi}, \mathbf{y})u_i(\mathbf{y}) \right] dS(\mathbf{y}) \qquad (1.27)$$

An alternative form of equation (1.27) is:

$$\frac{\partial u_k(\boldsymbol{\xi})}{\partial \xi_m} = \int_{\partial B} \left[\frac{\partial U_{ik}}{\partial \xi_m}(\boldsymbol{\xi}, \mathbf{y})\sigma_{ij}(\mathbf{y}) - \frac{\partial \Sigma_{ijk}}{\partial \xi_m}(\boldsymbol{\xi}, \mathbf{y})u_i(\mathbf{y}) \right] e_j \cdot d\mathbf{S}(\mathbf{y}) \qquad (1.28)$$

Stress components at an internal point $\boldsymbol{\xi}$ can be obtained from either of equations (1.27) or (1.28) by using Hooke's law:

$$\sigma_{ij} = \lambda u_{k,k}\delta_{ij} + \mu(u_{i,j} + u_{j,i}) \qquad (1.29)$$

It is sometimes convenient, however, to write the internal stress directly. This equation, corresponding (for example) to (1.27) is:

$$\sigma_{ij}(\boldsymbol{\xi}) = \int_{\partial B} [D_{ijk}(\boldsymbol{\xi}, \mathbf{y})\tau_k(\mathbf{y}) - S_{ijk}(\boldsymbol{\xi}, \mathbf{y})u_k(\mathbf{y})] dS(\mathbf{y}) \qquad (1.30)$$

where the new kernels $\mathbf{D}$ and $\mathbf{S}$ are:

$$D_{ijk} = E_{ijmn}\frac{\partial U_{km}}{\partial \xi_n} = \lambda\frac{\partial U_{km}}{\partial \xi_m}\delta_{ij} + \mu\left(\frac{\partial U_{ki}}{\partial \xi_j} + \frac{\partial U_{kj}}{\partial \xi_i}\right) = -\Sigma_{ijk} \qquad (1.31)$$

$$\begin{aligned}
S_{ijk} &= E_{ijmn}\frac{\partial \Sigma_{kpm}}{\partial \xi_n}n_p = \lambda\frac{\partial \Sigma_{kpm}}{\partial \xi_m}n_p\delta_{ij} + \mu\left(\frac{\partial \Sigma_{kpi}}{\partial \xi_j} + \frac{\partial \Sigma_{kpj}}{\partial \xi_i}\right)n_p \\
&= \frac{G}{4\pi(1-\nu)r^3}\left[3\frac{\partial r}{\partial n}\left[(1-2\nu)\delta_{ij}r_{,k} + \nu(\delta_{ik}r_{,j} + \delta_{jk}r_{,i}) - 5r_{,i}r_{,j}r_{,k}\right]\right] \\
&\quad + \frac{G}{4\pi(1-\nu)r^3}[3\nu(n_i r_{,j}r_{,k} + n_j r_{,i}r_{,k}) \\
&\quad + (1-2\nu)(3n_k r_{,i}r_{,j} + n_j\delta_{ik} + n_i\delta_{jk}) - (1-4\nu)n_k\delta_{ij}]
\end{aligned} \qquad (1.32)$$

Again, the normal $\mathbf{n}$ is defined at the (boundary) field point $\mathbf{y}$. Also:

$$\frac{\partial U_{ik}}{\partial \xi_m}(\boldsymbol{\xi}, \mathbf{y}) = -U_{ik,m} , \qquad \frac{\partial \Sigma_{ijk}}{\partial \xi_m}(\boldsymbol{\xi}, \mathbf{y}) = -\Sigma_{ijk,m} \qquad (1.33)$$

It is important to note that **D** becomes strongly singular, and **S** hypersingular as a source point approaches a field point (i.e. as $r \to 0$).

For future use in Chapter 4, it is useful to rewrite (1.28) using (1.33). This equation is:

$$u_{k,m}(\boldsymbol{\xi}) = -\int_{\partial B} [U_{ik,m}(\boldsymbol{\xi}, \mathbf{y})\sigma_{ij}(\mathbf{y}) - \Sigma_{ijk,m}(\boldsymbol{\xi}, \mathbf{y})u_i(\mathbf{y})] \, n_j(\mathbf{y}) dS(\mathbf{y}) \qquad (1.34)$$

Again, as one takes the limit $\boldsymbol{\xi} \to \mathbf{x}$ in any of the equations (1.27), (1.28) or (1.30), one must take the finite part of the corresponding right hand side (see Section 1.4.3). For example, (1.28) and (1.30) become, respectively:

$$\begin{aligned}
\frac{\partial u_k(\mathbf{x})}{\partial x_m} &= \lim_{\boldsymbol{\xi} \to \mathbf{x}} \int_{\partial B} \left[\frac{\partial U_{ik}}{\partial \xi_m}(\boldsymbol{\xi}, \mathbf{y})\sigma_{ij}(\mathbf{y}) - \frac{\partial \Sigma_{ijk}}{\partial \xi_m}(\boldsymbol{\xi}, \mathbf{y})u_i(\mathbf{y}) \right] n_j(\mathbf{y}) dS(\mathbf{y}) \\
&= \fint_{\partial B} \left[\frac{\partial U_{ik}}{\partial x_m}(\mathbf{x}, \mathbf{y})\sigma_{ij}(\mathbf{y}) - \frac{\partial \Sigma_{ijk}}{\partial x_m}(\mathbf{x}, \mathbf{y})u_i(\mathbf{y}) \right] n_j(\mathbf{y}) dS(\mathbf{y}) \quad (1.35)
\end{aligned}$$

$$\begin{aligned}
\sigma_{ij}(\mathbf{x}) &= \lim_{\boldsymbol{\xi} \to \mathbf{x}} \int_{\partial B} [D_{ijk}(\boldsymbol{\xi}, \mathbf{y})\tau_k(\mathbf{y}) - S_{ijk}(\boldsymbol{\xi}, \mathbf{y})u_k(\mathbf{y})] \, dS(\mathbf{y}) \\
&= \fint_{\partial B} [D_{ijk}(\mathbf{x}, \mathbf{y})\tau_k(\mathbf{y}) - S_{ijk}(\mathbf{x}, \mathbf{y})u_k(\mathbf{y})] \, dS(\mathbf{y}) \qquad (1.36)
\end{aligned}$$

Also, for future reference, one notes that the traction at a boundary point is:

$$\tau_i(\mathbf{x}) = n_j(\mathbf{x}) \lim_{\boldsymbol{\xi} \to \mathbf{x}} \int_{\partial B} [D_{ijk}(\boldsymbol{\xi}, \mathbf{y})\tau_k(\mathbf{y}) - S_{ijk}(\boldsymbol{\xi}, \mathbf{y})u_k(\mathbf{y})] \, dS(\mathbf{y}) \qquad (1.37)$$

Fully regularized forms of equations (1.35) and (1.36), that only contain weakly singular integrals, are available in the literature (see, for example, Cruse and Richardson [39]). These equations, that can be collocated at an irregular point $\mathbf{x} \in \partial B$ provided that the stress and displacement fields in (1.38, 1.39) satisfy certain smoothness requirements (see Martin et al. [93] and, also, Section 1.4.4 of this chapter) are:

$$\begin{aligned}
0 &= \int_{\partial B} U_{ik,m}(\mathbf{x}, \mathbf{y}) \left[\sigma_{ij}(\mathbf{y}) - \sigma_{ij}(\mathbf{x}) \right] n_j(\mathbf{y}) dS(\mathbf{y}) \\
&\quad - \int_{\partial B} \Sigma_{ijk,m}(\mathbf{x}, \mathbf{y}) \left[u_i(\mathbf{y}) - u_i(\mathbf{x}) - u_{i,\ell}(\mathbf{x})(y_\ell - x_\ell) \right] n_j(\mathbf{y}) dS(\mathbf{y}) \quad (1.38)
\end{aligned}$$

$$0 = \int_{\partial B} D_{ijk}(\mathbf{x}, \mathbf{y}) \left[\sigma_{kp}(\mathbf{y}) - \sigma_{kp}(\mathbf{x}) \right] n_p(\mathbf{y}) dS(\mathbf{y})$$

$$- \int_{\partial B} S_{ijk}(\mathbf{x}, \mathbf{y}) \left[u_k(\mathbf{y}) - u_k(\mathbf{x}) - u_{k,p}(\mathbf{x})(y_p - x_p) \right] dS(\mathbf{y}) \quad (1.39)$$

An alternate version of (1.39) that can only be collocated at a regular point $\mathbf{x} \in \partial B$ is:

$$0 = \int_{\partial B} D_{ijk}(\mathbf{x}, \mathbf{y})[\tau_k(\mathbf{y}) - \tau_k(\mathbf{x})] dS(\mathbf{y})$$

$$- \sigma_{km}(\mathbf{x}) \int_{\partial B} D_{ijk}(\mathbf{x}, \mathbf{y})(n_m(\mathbf{y}) - n_m(\mathbf{x})) dS(\mathbf{y})$$

$$- \int_{\partial B} S_{ijk}(\mathbf{x}, \mathbf{y}) \left[u_k(\mathbf{y}) - u_k(\mathbf{x}) - u_{k,m}(\mathbf{x})(y_m - x_m) \right] dS(\mathbf{y}) \quad (1.40)$$

Finally, taking the inner product of (1.40) with the normal at the source point gives:

$$0 = \int_{\partial B} D_{ijk}(\mathbf{x}, \mathbf{y})n_j(\mathbf{x})[\tau_k(\mathbf{y}) - \tau_k(\mathbf{x})] dS(\mathbf{y})$$

$$- \sigma_{km}(\mathbf{x}) \int_{\partial B} D_{ijk}(\mathbf{x}, \mathbf{y})n_j(\mathbf{x})[n_m(\mathbf{y}) - n_m(\mathbf{x})] dS(\mathbf{y})$$

$$- \int_{\partial B} S_{ijk}(\mathbf{x}, \mathbf{y})n_j(\mathbf{x}) \left[u_k(\mathbf{y}) - u_k(\mathbf{x}) - u_{k,m}(\mathbf{x})(y_m - x_m) \right] dS(\mathbf{y}) \quad (1.41)$$

1.2.2.1   Displacement gradient on the bounding surface

The gradient of the displacement $\mathbf{u}$ is required for the regularized HBIEs (1.38 - 1.41). Lutz et al. [89] have proposed a scheme for carrying this out. Details of this procedure are available in [27] and are given below.

The (right-handed) global Cartesian coordinates, as before, are (x_1, x_2, x_3). Consider (right-handed) local Cartesian coordinates (x'_1, x'_2, x'_3) at a regular point P on ∂B as shown in Figure 1.2. The local coordinate system is oriented such that the x'_1 and x'_2 coordinates lie along the tangential unit vectors $\mathbf{t}_1$ and $\mathbf{t}_2$ while x'_3 is measured along the outward normal unit vector $\mathbf{n}$ to ∂B as defined in equation (1.14).

Therefore, one has:

$$\mathbf{x'} = \mathbf{Q}\mathbf{x} \qquad (1.42)$$

$$\mathbf{u'} = \mathbf{Q}\mathbf{u} \qquad (1.43)$$

where u'_k, $k = 1, 2, 3$ are the components of the displacement vector $\mathbf{u}$ in the local coordinate frame, and the orthogonal transformation matrix $\mathbf{Q}$ has the components:

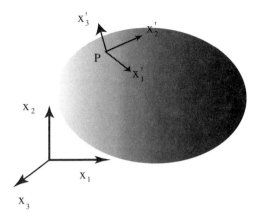

Figure 1.2: Local coordinate system on the surface of a body (from [27])

$$\mathbf{Q} = \begin{bmatrix} t_{11} & t_{12} & t_{13} \\ t_{21} & t_{22} & t_{23} \\ n_1 & n_2 & n_3 \end{bmatrix} \tag{1.44}$$

with t_{ij} the jth component of the ith unit tangent vector and (n_1, n_2, n_3) the components of the unit normal vector.

The tangential derivatives of the displacement, in local coordinates, are $u'_{i,k'}$, $i = 1, 2, 3$; $k = 1, 2$. These quantities are obtained as follows:

$$u'_{i,k'} \equiv \frac{\partial u'_i}{\partial s_k} = Q_{ij} \frac{\partial u_j}{\partial s_k} \tag{1.45}$$

where $\partial u'_i / \partial s_k$ are tangential derivatives of u_i at P with $s_1 = x'_1$ and $s_2 = x'_2$.

The remaining components of $\nabla \mathbf{u}$ in local coordinates are obtained from Hooke's law (see [89]) as:

$$\frac{\partial u'_1}{\partial x'_3} = \frac{\tau'_1}{G} - \frac{\partial u'_3}{\partial x'_1}$$

$$\frac{\partial u'_2}{\partial x'_3} = \frac{\tau'_2}{G} - \frac{\partial u'_3}{\partial x'_2}$$

$$\frac{\partial u'_3}{\partial x'_3} = \frac{(1-2\nu)\tau'_3}{2G(1-\nu)} - \frac{\nu}{1-\nu}\left[\frac{\partial u'_1}{\partial x'_1} + \frac{\partial u'_2}{\partial x'_2}\right] \tag{1.46}$$

where τ'_k, $k = 1, 2, 3$, are the components of the traction vector in local coordinates.

The components of the displacement gradient tensor, in the local coordinate system, are now known. They can be written as:

$$(\boldsymbol{\nabla}\mathbf{u})_{local} \equiv \mathbf{A}' = \begin{bmatrix} u'_{1,1'} & u'_{1,2'} & u'_{1,3'} \\ u'_{2,1'} & u'_{2,2'} & u'_{2,3'} \\ u'_{3,1'} & u'_{3,2'} & u'_{3,3'} \end{bmatrix} \tag{1.47}$$

Finally, the components of $\boldsymbol{\nabla}\mathbf{u}$ in the global coordinate frame are obtained from those in the local coordinate frame by using the tensor transformation rule:

$$(\boldsymbol{\nabla}\mathbf{u})_{global} \equiv \mathbf{A} = \mathbf{Q}^T \mathbf{A}' \mathbf{Q} = \begin{bmatrix} u_{1,1} & u_{1,2} & u_{1,3} \\ u_{2,1} & u_{2,2} & u_{2,3} \\ u_{3,1} & u_{3,2} & u_{3,3} \end{bmatrix} \tag{1.48}$$

The gradient of the displacement field in global coordinates is now ready for use in equations (1.38 - 1.41).

1.3   Nearly Singular Integrals in Linear Elasticity

It is well known that the first step in the BEM is to solve the primary problem on the bounding surface of a body (e.g. equation (1.25)) and obtain all the displacements and tractions on this surface. The next steps are to obtain the displacements and stresses at selected points inside a body, from equations such as (1.17) and (1.30). It has been known in the BEM community for many years, dating back to Cruse [37], that one experiences difficulties when trying to numerically evaluate displacements and stresses at points inside a body that are close to its bounding surface (the so-called near-singular or boundary layer problem). Various authors have addressed this issue over the last 3 decades. This section describes a new method recently proposed by Mukherjee et al. [104].

1.3.1   Displacements at Internal Points Close to the Boundary

The displacement at a point inside an elastic body can be determined from either of the (equivalent) equations (1.17) or (1.21). A continuous version of (1.21), from Cruse and Richardson [39] is:

$$u_k(\boldsymbol{\xi}) = u_k(\hat{\mathbf{x}}) + \int_{\partial B} [\, U_{ik}(\boldsymbol{\xi},\mathbf{y})\sigma_{ij}(\mathbf{y}) - \Sigma_{ijk}(\boldsymbol{\xi},\mathbf{y})\{u_i(\mathbf{y}) - u_i(\hat{\mathbf{x}})\} \,]\, n_j(\mathbf{y})dS(\mathbf{y})$$
$$\tag{1.49}$$

where $\boldsymbol{\xi} \in B$ is an internal point *close* to ∂B and a target point $\hat{\mathbf{x}} \in \partial B$ is *close* to the point $\boldsymbol{\xi}$ (see Fig. 1.3). An alternative form of (1.49) is:

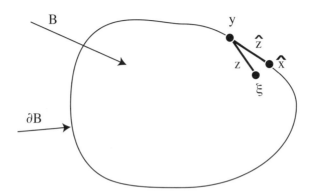

Figure 1.3: A body with source point $\boldsymbol{\xi}$, field point $\mathbf{y}$ and target point $\hat{\mathbf{x}}$ (from [104])

$$u_k(\boldsymbol{\xi}) = u_k(\hat{\mathbf{x}}) + \int_{\partial B} [\, U_{ik}(\boldsymbol{\xi}, \mathbf{y})\tau_i(\mathbf{y}) - T_{ik}(\boldsymbol{\xi}, \mathbf{y})\{u_i(\mathbf{y}) - u_i(\hat{\mathbf{x}})\} \,]\, dS(\mathbf{y}) \quad (1.50)$$

Equation (1.49) (or (1.50)) is called "continuous" since it has a continuous limit to the boundary (LTB as $\boldsymbol{\xi} \to \hat{\mathbf{x}} \in \partial B$) provided that $u_i(\mathbf{y}) \in C^{0,\alpha}$ (i.e. Hölder continuous). Taking this limit is the standard approach for obtaining the well-known regularized form (1.26) (or (1.25)).

In this work, however, equation (1.49) (or (1.50)) is put to a different, and novel use. It is first observed that T_{ik} in equation (1.50) is $\mathcal{O}(1/r^2(\boldsymbol{\xi},\mathbf{y}))$ as $\boldsymbol{\xi} \to \mathbf{y}$, whereas $\{u_i(\mathbf{y}) - u_i(\hat{\mathbf{x}})\}$ is $\mathcal{O}(r(\hat{\mathbf{x}},\mathbf{y}))$ as $\mathbf{y} \to \hat{\mathbf{x}}$. Therefore, as $\mathbf{y} \to \hat{\mathbf{x}}$, the product $T_{ik}(\boldsymbol{\xi},\mathbf{y})\{u_i(\mathbf{y}) - u_i(\hat{\mathbf{x}})\}$, which is $\mathcal{O}(r(\hat{\mathbf{x}},\mathbf{y})/r^2(\boldsymbol{\xi},\mathbf{y}))$, $\to 0$! As a result, equation (1.50) (or (1.49)) can be used to easily and accurately evaluate the displacement components $u_k(\boldsymbol{\xi})$ for $\boldsymbol{\xi} \in B$ close to ∂B. *This idea is the main contribution of [104]*.

It is noted here that while it is usual to use (1.17) (or (1.21)) to evaluate $u_k(\boldsymbol{\xi})$ when $\boldsymbol{\xi}$ is *far* from ∂B, equation (1.49) (or (1.50)) is also valid in this case. (The target point $\hat{\mathbf{x}}$ can be chosen as *any* point on ∂B when $\boldsymbol{\xi}$ is far from ∂B). Therefore, it is advisable to use the continuous equation (1.49) (or (1.50)) universally for all points $\boldsymbol{\xi} \in B$. This procedure would eliminate the need to classify, a priori, whether $\boldsymbol{\xi}$ is near to, or far from ∂B.

1.3.2   Stresses at Internal Points Close to the Boundary

The displacement gradient at a point $\boldsymbol{\xi} \in B$ can be obtained from equation (1.34) or the stress at this point from (1.30). Continuous versions of (1.34) and (1.30) can be written as [39]:

$$u_{k,n}(\boldsymbol{\xi}) \;=\; u_{k,n}(\hat{\mathbf{x}}) - \int_{\partial B} U_{ik,n}(\boldsymbol{\xi}, \mathbf{y}) \left[\sigma_{ij}(\mathbf{y}) - \sigma_{ij}(\hat{\mathbf{x}})\right] n_j(\mathbf{y}) dS(\mathbf{y})$$

$$+ \int_{\partial B} \Sigma_{ijk,n}(\boldsymbol{\xi}, \mathbf{y}) \left[u_i(\mathbf{y}) - u_i(\hat{\mathbf{x}}) - u_{i,\ell}(\hat{\mathbf{x}}) (y_\ell - \hat{x}_\ell)\right] n_j(\mathbf{y}) dS(\mathbf{y}) \qquad (1.51)$$

$$\sigma_{ij}(\boldsymbol{\xi}) \;=\; \sigma_{ij}(\hat{\mathbf{x}}) + \int_{\partial B} D_{ijk}(\boldsymbol{\xi}, \mathbf{y}) [\tau_k(\mathbf{y}) - \sigma_{km}(\hat{\mathbf{x}}) n_m(\mathbf{y})] dS(\mathbf{y})$$

$$- \int_{\partial B} S_{ijk}(\boldsymbol{\xi}, \mathbf{y}) [u_k(\mathbf{y}) - u_k(\hat{\mathbf{x}}) - u_{k,\ell}(\hat{\mathbf{x}})(y_\ell - \hat{x}_\ell)] dS(\mathbf{y}) \qquad (1.52)$$

The integrands in equations (1.51) (or (1.52)) are $\mathcal{O}(r(\hat{\mathbf{x}}, \mathbf{y})/r^2(\boldsymbol{\xi}, \mathbf{y}))$ and $\mathcal{O}(r^2(\hat{\mathbf{x}}, \mathbf{y})/r^3(\boldsymbol{\xi}, \mathbf{y}))$ as $\mathbf{y} \to \hat{\mathbf{x}}$. Similar to the behavior of the continuous BIEs in the previous subsection, the integrands in equations (1.51) and (1.52) $\to 0$ as $\mathbf{y} \to \hat{\mathbf{x}}$. Either of these equations, therefore, is very useful for evaluating the stresses at an internal point $\boldsymbol{\xi}$ that is *close* to ∂B. Of course (please see the discussion regarding displacements in the previous section), they can also be conveniently used to evaluate displacement gradients or stresses at *any* point $\boldsymbol{\xi} \in B$.

Henceforth, use of equations (1.17), (1.21), (1.30) or (1.34) will be referred to as the standard method, while use of equations (1.49), (1.50), (1.51) or (1.52) will be referred to as the new method.

1.4   Finite Parts of Hypersingular Equations

A discussion of finite parts (FPs) of hypersingular BIEs (see e.g. equations (1.9 -1.11)) is the subject of this section. The general theory of finite parts is presented first. This is followed by applications of the theory in potential theory and in linear elasticity. Further details are available in Mukherjee [102].

1.4.1   Finite Part of a Hypersingular Integral Collocated at an Irregular Boundary Point

1.4.1.1   Definition

Consider, for specificity, the space $\mathbf{R}^3$, and let S be a surface in $\mathbf{R}^3$. Let the points $\mathbf{x} \in S$ and $\boldsymbol{\xi} \notin S$. Also, let $\hat{S}$ and $\bar{S} \subset \hat{S}$ be two neighborhoods (in S) of $\mathbf{x}$ such that $\mathbf{x} \in \bar{S}$ (Figure 1.4). The point $\mathbf{x}$ can be an irregular point on S.

Let the function $K(\mathbf{x}, \mathbf{y})$, $\mathbf{y} \in S$, have its only singularity at $\mathbf{x} = \mathbf{y}$ of the form $1/r^3$ where $r = |\mathbf{x} - \mathbf{y}|$, and let $\phi(\mathbf{y})$ be a function that has no singularity in S and is of class $C^{1,\alpha}$ at $\mathbf{y} = \mathbf{x}$ for some $\alpha > 0$.

The finite part of the integral

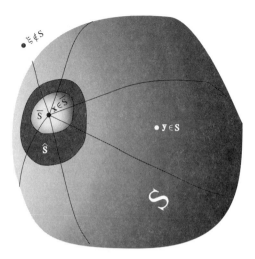

Figure 1.4: A surface S with regions $\hat{S}$ and $\bar{S}$ and points $\boldsymbol{\xi}, \mathbf{x}$ and $\mathbf{y}$ (from [102])

$$I(\mathbf{x}) = \int_S K(\mathbf{x}, \mathbf{y})\phi(\mathbf{y})dS(\mathbf{y}) \tag{1.53}$$

is defined as:

$$
\begin{aligned}
\fint_S K(\mathbf{x}, \mathbf{y})\phi(\mathbf{y})dS(\mathbf{y}) = {} & \int_{S\backslash\hat{S}} K(\mathbf{x}, \mathbf{y})\phi(\mathbf{y})dS(\mathbf{y}) \\
& + \int_{\hat{S}} K(\mathbf{x}, \mathbf{y})[\phi(\mathbf{y}) - \phi(\mathbf{x}) - \phi_{,p}(\mathbf{x})(y_p - x_p)]dS(\mathbf{y}) \\
& + \phi(\mathbf{x})A(\hat{S}) + \phi_{,p}(\mathbf{x})B_p(\hat{S}) \tag{1.54}
\end{aligned}
$$

where $\hat{S}$ is any arbitrary neighborhood (in S) of $\mathbf{x}$ and:

$$A(\hat{S}) = \fint_{\hat{S}} K(\mathbf{x}, \mathbf{y})dS(\mathbf{y}) \tag{1.55}$$

$$B_p(\hat{S}) = \fint_{\hat{S}} K(\mathbf{x}, \mathbf{y})(y_p - x_p)dS(\mathbf{y}) \tag{1.56}$$

The above FP definition can be easily extended to any number of physical dimensions and any order of singularity of the kernel function $K(\mathbf{x}, \mathbf{y})$. Please refer to Toh and Mukherjee [168] for further discussion of a previous closely related FP definition for the case when $\mathbf{x}$ is a regular point on S, and to Mukherjee [101] for a discussion of the relationship of this FP to the CPV of an integral when its CPV exists.

1.4.1.2   Evaluation of A and $\mathbf{B}$

There are several equivalent ways for evaluating A and $\mathbf{B}$.

Method one. Replace S by $\hat{S}$ and $\hat{S}$ by $\bar{S}$ in equation (1.54). Now, setting $\phi(\mathbf{y}) = 1$ in (1.54) and using (1.55), one gets:

$$A(\hat{S}) - A(\bar{S}) = \int_{\hat{S}\backslash\bar{S}} K(\mathbf{x},\mathbf{y})dS(\mathbf{y}) \qquad (1.57)$$

Next, setting $\phi(\mathbf{y}) = (y_p - x_p)$ (note that, in this case, $\phi(\mathbf{x}) = 0$ and $\phi_{,p}(\mathbf{x}) = 1$) in (1.54), and using (1.56), one gets:

$$B_p(\hat{S}) - B_p(\bar{S}) = \int_{\hat{S}\backslash\bar{S}} K(\mathbf{x},\mathbf{y})(y_p - x_p)dS(\mathbf{y}) \qquad (1.58)$$

The formulae (1.57) and (1.58) are most useful for obtaining A and $\mathbf{B}$ when $\hat{S}$ is an open surface and Stoke regularization is employed. An example is the application of the FP definition (1.54) (for a regular collocation point) in Toh and Mukherjee [168], to regularize a hypersingular integral that appears in the HBIE formulation for the scattering of acoustic waves by a thin scatterer. The resulting regularized equation is shown in [168] to be equivalent to the result of Krishnasamy et al. [75]. Equations (1.57) and (1.58) are also used in Mukherjee and Mukherjee [99] and in Section 3.2 of [102].

Method two. From equation (1.57):

$$A(\hat{S}) - A(\bar{S}) = \int_{\hat{S}\backslash\bar{S}} K(\mathbf{x},\mathbf{y})dS(\mathbf{y}) = \lim_{\boldsymbol{\xi}\to\mathbf{x}} \int_{\hat{S}\backslash\bar{S}} K(\boldsymbol{\xi},\mathbf{y})dS(\mathbf{y}) \qquad (1.59)$$

The second equality above holds since $K(\mathbf{x},\mathbf{y})$ is regular for $\mathbf{x} \in \bar{S}$ and $\mathbf{y} \in \hat{S}\backslash\bar{S}$. Assuming that the limits:

$$\lim_{\boldsymbol{\xi}\to\mathbf{x}} \int_{\hat{S}} K(\boldsymbol{\xi},\mathbf{y})dS(\mathbf{y}), \quad \lim_{\boldsymbol{\xi}\to\mathbf{x}} \int_{\bar{S}} K(\boldsymbol{\xi},\mathbf{y})dS(\mathbf{y})$$

exist, then:

$$A(\hat{S}) = \lim_{\boldsymbol{\xi}\to\mathbf{x}} \int_{\hat{S}} K(\boldsymbol{\xi},\mathbf{y})dS(\mathbf{y}) \qquad (1.60)$$

Similarly:

$$B_p(\hat{S}) = \lim_{\boldsymbol{\xi}\to\mathbf{x}} \int_{\hat{S}} K(\boldsymbol{\xi},\mathbf{y})(y_p - x_p)dS(\mathbf{y}) \qquad (1.61)$$

Equations (1.60) and (1.61) are most useful for evaluating A and $\mathbf{B}$ when $\hat{S} = \partial B$, a closed surface that is the entire boundary of a body B. Examples appear in Sections 1.4.2 and 1.4.3 of this chapter.

Method three. A third way for evaluation of A and **B** is to use an auxiliary surface (or "tent") as first proposed for fracture mechanics analysis by Lutz et al. [89]. (see, also, Mukherjee et al. [110], Mukherjee [105] and Section 3.2.1 of [102]. This method is useful if S is an open surface.

1.4.1.3 The FP and the LTB

There is a very simple connection between the FP, defined above, and the LTB approach employed by Gray and his coauthors. With, as before, $\boldsymbol{\xi} \notin S$, $\mathbf{x} \in S$ (**x** can be an irregular point on S), $K(\mathbf{x}, \mathbf{y}) = \mathcal{O}(|\mathbf{x} - \mathbf{y}|^{-3})$ as $\mathbf{y} \to \mathbf{x}$ and $\phi(\mathbf{y}) \in C^{1,\alpha}$ at $\mathbf{y} = \mathbf{x}$, this can be stated as:

$$\lim_{\boldsymbol{\xi} \to \mathbf{x}} \int_S K(\boldsymbol{\xi}, \mathbf{y})\phi(\mathbf{y})dS(\mathbf{y}) = \fint_S K(\mathbf{x}, \mathbf{y})\phi(\mathbf{y})dS(\mathbf{y}) \tag{1.62}$$

Of course, $\boldsymbol{\xi}$ can approach **x** from either side of S.

Proof of equation (1.62). Consider the first and second terms on the right-hand side of equation (1.54). Since these integrands are regular in their respective domains of integration, one has:

$$\int_{S\backslash\hat{S}} K(\mathbf{x}, \mathbf{y})\phi(\mathbf{y})dS(\mathbf{y}) = \lim_{\boldsymbol{\xi} \to \mathbf{x}} \int_{S\backslash\hat{S}} K(\boldsymbol{\xi}, \mathbf{y})\phi(\mathbf{y})dS(\mathbf{y}) \tag{1.63}$$

and

$$\int_{\hat{S}} K(\mathbf{x}, \mathbf{y})[\phi(\mathbf{y}) - \phi(\mathbf{x}) - \phi_{,p}(\mathbf{x})\ (y_p - x_p)]dS(\mathbf{y})$$
$$= \lim_{\boldsymbol{\xi} \to \mathbf{x}} \int_{\hat{S}} K(\boldsymbol{\xi}, \mathbf{y})[\phi(\mathbf{y}) - \phi(\boldsymbol{\xi}) - \phi_{,p}(\boldsymbol{\xi})(y_p - \xi_p)]dS(\mathbf{y}) \tag{1.64}$$

Use of equations (1.60, 1.61, 1.63 and 1.64) in (1.54) proves equation (1.62).

1.4.2 Gradient BIE for 3-D Laplace's Equation

This section is concerned with an application of equation (1.54) for collocation of the HBIE (1.9), for the 3-D Laplace equation, at an irregular boundary point. A complete exclusion zone, $\hat{S} = \partial B$ is used here. An application of a vanishing exclusion zone, for collocation of the HBIE for the 2-D Laplace equation, at an irregular boundary point, is presented in Mukherjee [102].

Using equations (1.4) and (1.6), equations (1.9) and (1.10) are first written in the slightly different equivalent forms:

$$\frac{\partial u(\boldsymbol{\xi})}{\partial \xi_i} = \int_{\partial B} [D_i(\boldsymbol{\xi}, \mathbf{y})\tau(\mathbf{y}) - S_i(\boldsymbol{\xi}, \mathbf{y})u(\mathbf{y})]\, dS(\mathbf{y}) \tag{1.65}$$

$$\frac{\partial u(\mathbf{x})}{\partial x_i} = \fint_{\partial B} [D_i(\mathbf{x}, \mathbf{y})\tau(\mathbf{y}) - S_i(\mathbf{x}, \mathbf{y})u(\mathbf{y})]\, dS(\mathbf{y}) \tag{1.66}$$

where:

$$D_i(\mathbf{x}, \mathbf{y}) = -G_{,i}(\mathbf{x}, \mathbf{y}) , \quad S_i = -H_{k,i}(\mathbf{x}, \mathbf{y})n_k(\mathbf{y}) \tag{1.67}$$

Use of (1.54) in (1.66), with $S = \hat{S} = \partial B$, results in:

$$
\begin{aligned}
u_{,i}(\mathbf{x}) = {} & \int_{\partial B} D_i(\mathbf{x}, \mathbf{y})\Big[u_{,p}(\mathbf{y}) - u_{,p}(\mathbf{x})\Big]n_p(\mathbf{y})dS(\mathbf{y}) \\
& - \int_{\partial B} S_i(\mathbf{x}, \mathbf{y})\Big[u(\mathbf{y}) - u(\mathbf{x}) - u_{,p}(\mathbf{x})(y_p - x_p)\Big]dS(\mathbf{y}) \\
& - A_i(\partial B)u(\mathbf{x}) + C_{ip}(\partial B)u_{,p}(\mathbf{x})
\end{aligned}
\tag{1.68}
$$

where, using method two in Section 1.4.1.2:

$$A_i(\partial B) = \lim_{\boldsymbol{\xi} \to \mathbf{x}} \int_{\partial B} S_i(\boldsymbol{\xi}, \mathbf{y})dS(\mathbf{y}) \tag{1.69}$$

$$C_{ip}(\partial B) = \lim_{\boldsymbol{\xi} \to \mathbf{x}} \int_{\partial B} [D_i(\boldsymbol{\xi}, \mathbf{y})n_p(\mathbf{y}) - S_i(\boldsymbol{\xi}, \mathbf{y})(y_p - \xi_p)]\, dS(\mathbf{y}) \tag{1.70}$$

It is noted here that the (possibly irregular) boundary point $\mathbf{x}$ is approached from $\boldsymbol{\xi} \in B$, i.e. from inside the body B.

The quantities $\mathbf{A}$ and $\mathbf{C}$ can be easily evaluated using the imposition of simple solutions. Following Rudolphi [143], use of the uniform solution $u(\mathbf{y}) = c$ (c is a constant) in equation (1.65) gives:

$$\int_{\partial B} S_i(\boldsymbol{\xi}, \mathbf{y})dS(\mathbf{y}) = 0 \tag{1.71}$$

while use of the linear solution:

$$
\begin{aligned}
u &= u(\boldsymbol{\xi}) + (y_p - \xi_p)u_{,p}(\boldsymbol{\xi}) \\
\tau(\mathbf{y}) &= \frac{\partial u}{\partial y_k}n_k(\mathbf{y}) = u_{,p}(\boldsymbol{\xi})n_p(\mathbf{y}) \qquad \text{(with } p = 1, 2, 3\text{)}
\end{aligned}
\tag{1.72}
$$

in equation (1.65) (together with (1.71)) gives:

$$\int_{\partial B} [D_i(\boldsymbol{\xi}, \mathbf{y})n_p(\mathbf{y}) - S_i(\boldsymbol{\xi}, \mathbf{y})(y_p - \xi_p)]\, dS(\mathbf{y}) = \delta_{ip} \tag{1.73}$$

Therefore, (assuming continuity) $A_i(\partial B) = 0$, $C_{ip}(\partial B) = \delta_{ip}$, and (1.68) yields a simple, fully regularized form of (1.66) as:

$$
\begin{aligned}
0 = {} & \int_{\partial B} D_i(\mathbf{x}, \mathbf{y})[u_{,p}(\mathbf{y}) - u_{,p}(\mathbf{x})]n_p(\mathbf{y})dS(\mathbf{y}) \\
& - \int_{\partial B} S_i(\mathbf{x}, \mathbf{y})[u(\mathbf{y}) - u(\mathbf{x}) - u_{,p}(\mathbf{x})(y_p - x_p)]dS(\mathbf{y})
\end{aligned}
\tag{1.74}
$$

which is equivalent to equation (1.11).

A few comments are in order. First, equation (1.74) is the same as Rudolphi's [143] equation (20) with (his) $\kappa = 1$ and (his) S_0 set equal to S and renamed ∂B. (See, also, Kane [68], equation (17.34)). Second, this equation can also be shown to be valid for the case $\boldsymbol{\xi} \notin B$, i.e. for an outside approach to the boundary point $\mathbf{x}$. Third, as noted before, $\mathbf{x}$ can be an edge or corner point on ∂B (provided, of course, that $u(\mathbf{y}) \in C^{1,\alpha}$ at $\mathbf{y} = \mathbf{x}$ - Rudolphi had only considered a regular boundary collocation point in his excellent paper that was published in 1991). Finally, as discussed in the Section 1.4.3, equation (1.74) is analogous to the regularized stress BIE in linear elasticity - equation (28) in Cruse and Richardson [39] .

1.4.3 Stress BIE for 3-D Elasticity

This section presents a proof of the fact that equation (1.39) is a regularized version of (1.36), valid at an irregular point $\mathbf{x} \in \partial B$, provided that the stress and displacement fields in (1.39) satisfy certain smoothness requirements. These smoothness requirements are discussed in Section 1.4.4. The approach is very similar to that used in Section 1.4.2.

The first step is to apply the FP equation (1.54) to regularize (1.36). With $S = \hat{S} = \partial B$, the result is:

$$
\begin{aligned}
\sigma_{ij}(\mathbf{x}) \;=\; & \int_{\partial B} D_{ijk}(\mathbf{x},\mathbf{y})\left[\sigma_{kp}(\mathbf{y}) - \sigma_{kp}(\mathbf{x})\right] n_p(\mathbf{y})dS(\mathbf{y}) \\
& - \int_{\partial B} S_{ijk}(\mathbf{x},\mathbf{y})\left[u_k(\mathbf{y}) - u_k(\mathbf{x}) - u_{k,p}(\mathbf{x})(y_p - x_p)\right]dS(\mathbf{y}) \\
& - A_{ijk}(\partial B)u_k(\mathbf{x}) + C_{ijkp}(\partial B)u_{k,p}(\mathbf{x}) \qquad (1.75)
\end{aligned}
$$

where, using method two in Section 1.4.1.2:

$$
A_{ijk}(\partial B) = \lim_{\boldsymbol{\xi}\to\mathbf{x}} \int_{\partial B} S_{ijk}(\boldsymbol{\xi},\mathbf{y})dS(\mathbf{y}) \qquad (1.76)
$$

$$
\begin{aligned}
C_{ijkp}(\partial B) \;=\; & \lim_{\boldsymbol{\xi}\to\mathbf{x}} \int_{\partial B} E_{m\ell kp}D_{ijm}(\boldsymbol{\xi},\mathbf{y})n_\ell(\mathbf{y})dS(\mathbf{y}) \\
& - \lim_{\boldsymbol{\xi}\to\mathbf{x}} \int_{\partial B} S_{ijk}(\boldsymbol{\xi},\mathbf{y})(y_p - \xi_p)dS(\mathbf{y}) \qquad (1.77)
\end{aligned}
$$

with $\mathbf{E}$ the elasticity tensor (see (1.23)) which appears in Hooke's law:

$$
\sigma_{m\ell} = E_{m\ell kp}u_{k,p} \qquad (1.78)
$$

Simple (rigid body and linear) solutions in linear elasticity (see, for example, Lutz et al. [89], Cruse and Richardson [39]) are now used in order to determine the quantities $\mathbf{A}$ and $\mathbf{C}$. Using the rigid body mode $u_k = c_k$ (c_k are arbitrary constants) in (1.30), one has:

$$0 = \int_{\partial B} S_{ijk}(\boldsymbol{\xi}, \mathbf{y}) dS(\mathbf{y}) \qquad (1.79)$$

while, using the linear solution:

$$
\begin{aligned}
u_k(\mathbf{y}) &= (y_p - \xi_p) u_{k,p}(\boldsymbol{\xi}), \quad u_{k,m}(\mathbf{y}) = u_{k,m}(\boldsymbol{\xi}), \\
\tau_k(\mathbf{y}) &= \sigma_{km}(\mathbf{y}) n_m(\mathbf{y}) = E_{kmrs} u_{r,s}(\boldsymbol{\xi}) n_m(\mathbf{y})
\end{aligned}
\qquad (1.80)
$$

in equation (1.30) gives:

$$\sigma_{ij}(\boldsymbol{\xi}) = u_{k,p}(\boldsymbol{\xi}) \int_{\partial B} \left[E_{m\ell k p} D_{ijm}(\boldsymbol{\xi}, \mathbf{y}) n_\ell(\mathbf{y}) - S_{ijk}(\boldsymbol{\xi}, \mathbf{y})(y_p - \xi_p) \right] dS(\mathbf{y})$$

$$(1.81)$$

Taking the limit $\boldsymbol{\xi} \to \mathbf{x}$ of (1.79), using continuity of the integral and comparing with (1.76), gives $\mathbf{A} = 0$. Taking the limit $\boldsymbol{\xi} \to \mathbf{x}$ of (1.81) and comparing with (1.77), one has:

$$\sigma_{ij}(\mathbf{x}) = C_{ijkp} u_{k,p}(\mathbf{x}) \qquad (1.82)$$

Comparing (1.82) with (1.78) yields $\mathbf{C}(\partial B) = \mathbf{E}$.

Therefore, equation (1.75) reduces to the simple regularized equation (1.39).

Equation (1.39) is equation (28) of Cruse and Richardson [39] in the present notation. As is the case in the present work, Cruse and Richardson [39] have also proved that their equation (28) is valid at a corner point, provided that the stress is continuous there.

It has been proved in this section that the regularized stress BIE (28) of Cruse and Richardson [39] can also be obtained from the FP definition (1.54) with a complete exclusion zone.

1.4.4 Solution Strategy for a HBIE Collocated at an Irregular Boundary Point

Hypersingular BIEs for a body B with boundary ∂B are considered here. Regularized HBIEs, obtained by using complete exclusion zones, e.g. equation (1.74) for potential theory or (1.39) for linear elasticity, are recommended as starting points.

An irregular collocation point $\mathbf{x}$ for 3-D problems is considered next. Let ∂B_n, $(n = 1, 2, 3, ..., N)$ be smooth pieces of ∂B that meet at an irregular point $\mathbf{x} \in \partial B$. Also, as before, let a source point, with coordinates x_k, be denoted by P, and a field point, with coordinates y_k, be denoted by Q.

Martin et al. [93] state the following requirements for collocating a regularized HBIE, such as (1.39) at an irregular point $P \in \partial B$. These are:

(i) The displacement **u** must satisfy the equilibrium equations in B.
(ii) (a) The stress $\boldsymbol{\sigma}$ must be continuous in B.
    (b) The stress $\boldsymbol{\sigma}$ must be continuous on ∂B.
(iii) $|u_i(Q_n) - u_i{}^L(Q_n; P)| = \mathcal{O}(r_n^{(1+\alpha)})$ as $r_n \to 0$,   for each n.
(iv) $[\sigma_{ij}(Q_n) - \sigma_{ij}(P)]n_j(Q_n) = \mathcal{O}(r_n^\alpha)$ as $r_n \to 0$,   for each n.

Box 1.1 Requirements for collocation of a HBIE at an irregular point (from [93]).

In the above, $r_n = |\mathbf{y}(Q_n) - \mathbf{x}(P)|$, $Q_n \in \partial B_n$, and $\alpha > 0$. Also,

$$u_i^L(Q_n; P) = u_i(P) + u_{i,j}(P)[y_j(Q_n) - x_j(P)] \tag{1.83}$$

There are two important issues to consider here.

The first is that, if there is to be any hope for collocating (1.39) at an irregular point P, the exact solution of a boundary value problem must satisfy conditions (i-iv) in Box 1.1. Clearly, one should not attempt this collocation if, for example, the stress is unbounded at P (this can easily happen - see an exhaustive study on the subject in Glushkov et al. [50]), or is bounded but discontinuous at P (e.g. at the tip of a wedge - see, for example, Zhang and Mukherjee [183]). The discussion in the rest of this book is limited to the class of problems, *referred to as the admissible class*, whose exact solutions satisfy conditions (i - iv).

The second issue refers to smoothness requirements on the interpolation functions for **u**, $\boldsymbol{\sigma}$ and the traction $\boldsymbol{\tau} = \mathbf{n} \cdot \boldsymbol{\sigma}$ in (1.39). It has proved very difficult, in practice, to find BEM interpolation functions that satisfy, a priori, (ii(b)-(iv)) in Box 1.1, for collocation at an irregular surface point on a 3-D body [93]. It has recently been proved in Mukherjee and Mukherjee [111], however, that interpolation functions used in the boundary contour method (BCM - see, for example, Mukherjee et al. [109], Mukherjee and Mukherjee [99]) satisfy these conditions a priori. Another important advantage of using these interpolation functions is that $\nabla \mathbf{u}$ can be directly computed from them at an irregular boundary point [99], without the need to use the (undefined) normal and tangent vectors at this point. In principle, these BCM interpolation functions can also be used in the BEM.

The BCM and the hypersingular BCM (HBCM) are discussed in detail in Chapter 4 of this book. Numerical results from the hypersingular BCM, collocated on edges and at corners, from Mukherjee and Mukherjee [111], are available in Chapter 4.

Chapter 2

ERROR ESTIMATION

Pointwise (i.e. that the error is evaluated at selected points) residual-based error estimates for Dirichlet, Neumann and mixed boundary value problems (BVPs) in linear elasticity are presented first in this chapter. Interesting relationships between the actual error and the hypersingular residuals are proved for the first two classes of problems, while heuristic error estimators are presented for mixed BVPs. Element-based error indicators, relying on the pointwise error measures presented earlier, are proposed next. Numerical results for two mixed BVPs in 2-D linear elasticity complete this chapter. Further details are available in [127].

2.1 Linear Operators

Boundary integral equations can be analyzed by viewing them as linear equations in a Hilbert space. A very readable account of this topic is available in Kress [73]. Following Sloan [155], it is assumed here that the boundary ∂B is a C^1 continuous closed Jordan curve given by the mapping:

$$z : [0,1] \to \partial B, \ z \in C^1, \ |z'| \neq 0$$

where $z \in \mathbf{C}$, the space of complex numbers. The present analysis excludes domains with corners. It is also assumed that any integrable function v on ∂B may be represented in a Fourier series:

$$v \sim \sum_{k=-\infty}^{\infty} \hat{v}(k) e^{2\pi i k x_1} = a_0 + \sum_{k=1}^{\infty} \left(a_k \cos(2\pi k x_1) + b_k \sin(2\pi k x_1) \right) \qquad (2.1)$$

where $i \equiv \sqrt{-1}$ and:

$$\hat{v}(k) = \int_0^1 e^{-2\pi i k x_1} v(x_1) dx_1, \qquad k \in \mathbf{Z} \qquad (2.2)$$

in which **Z** denotes the space of integers.

The following Lemma is very useful for the work presented in this chapter.

Lemma 1.   *If $\mathcal{A} : B_1 \to B_2$ is a continuous linear operator that has a continuous inverse, and $\mathcal{A}x = y$, then there exist real positive constants C_1 and C_2, such that:*

$$C_1 \|y\|_{B_2} \leq \|x\|_{B_1} \leq C_2 \|y\|_{B_2}$$

where $\|\cdot\|_{B_i}$ denotes a suitable norm of the appropriate function (in the Banach space B_i).

   Proof: The linearity and continuity of $\mathcal{A}$ and $\mathcal{A}^{-1}$ imply that $\|\mathcal{A}\|$ and $\|\mathcal{A}^{-1}\|$ are finite. From the Cauchy-Schwarz inequality, one has:

$$\|y\| = \|\mathcal{A}x\| \leq \|\mathcal{A}\|\|x\|$$

$$\|x\| = \|\mathcal{A}^{-1}y\| \leq \|\mathcal{A}^{-1}\|\|y\|$$

The result now follows by choosing $C_1 = 1/\|\mathcal{A}\|$ and $C_2 = \|\mathcal{A}^{-1}\|$. ◇

Returning to the problem at hand, the following operators are defined as:

$$(\mathcal{U}_{ij}\,v_j)(\boldsymbol{\xi}) := \int_{\partial B} U_{ij}(\boldsymbol{\xi},\mathbf{y})v_j(\mathbf{y})dS(\mathbf{y}) \tag{2.3}$$

$$(\mathcal{T}_{ij}\,v_j)(\boldsymbol{\xi}) := \int_{\partial B} T_{ij}^T(\boldsymbol{\xi},\mathbf{y})v_j(\mathbf{y})dS(\mathbf{y}) \tag{2.4}$$

$$(\mathcal{D}_{ijk}v_k)(\boldsymbol{\xi}) := \int_{\partial B} D_{ijk}(\boldsymbol{\xi},\mathbf{y})v_k(\mathbf{y})ds(\mathbf{y}) \tag{2.5}$$

$$(\mathcal{S}_{ijk}v_k)(\boldsymbol{\xi}) := \int_{\partial B} S_{ijk}(\boldsymbol{\xi},\mathbf{y})v_k(\mathbf{y})ds(\mathbf{y}) \tag{2.6}$$

$$(\mathcal{D}_{ik}^{(N)}\,v_k)(\boldsymbol{\xi},\mathbf{x}) := n_j(\mathbf{x})\int_{\partial B} D_{ijk}(\boldsymbol{\xi},\mathbf{y})v_k(\mathbf{y})ds(\mathbf{y}) \tag{2.7}$$

$$(\mathcal{S}_{ik}^{(N)}\,v_k)(\boldsymbol{\xi},\mathbf{x}) := n_j(\mathbf{x})\int_{\partial B} S_{ijk}(\boldsymbol{\xi},\mathbf{y})v_k(\mathbf{y})ds(\mathbf{y}) \tag{2.8}$$

   The operator $\mathcal{U}_{ij}$ is continuous onto the boundary, whereas $\mathcal{T}_{ij}$ and $\mathcal{D}_{ij}^{(N)}$ are not continuous (Tanaka et al. [162]) and give rise to additional bounded free terms in the limit. The hypersingular operator $\mathcal{S}_{ij}^{(N)}$ gives rise to unbounded terms that vanish when the integral is considered, for example, in the LTB sense. These terms depend on the smoothness of the boundary at the source point $\mathbf{x}$. In this work, the HBIE is collocated only at regular boundary points (where the boundary is locally smooth) inside boundary elements.

Using the operators defined above, the BIE (1.24) and HBIE (1.37) become, respectively:

$$\text{BIE}: \qquad u_i = \mathcal{U}_{ij}\tau_j - \mathcal{T}_{ij}u_j \qquad\qquad (2.9)$$

$$\text{HBIE}: \qquad \tau_i = \mathcal{D}_{ij}^{(N)}\tau_j - \mathcal{S}_{ij}^{(N)}u_j \qquad\qquad (2.10)$$

As in the case of potential theory (Menon et al. [96]), the LTB of the above integral operators has been used to obtain the singular integral equations (2.9) and (2.10).

Remark 1 *One should note, however, that key properties of the operators, such as continuity and invertibility, assume a certain regularity of the boundary (for instance no corners or cusps) [155]. These assumptions, of course, are too restrictive for the solution of practical engineering problems. Such assumptions have, nevertheless, been made here in order to obtain some mathematical understanding of the error estimation process that is described in Section 2.2. The numerical example problems do contain corners. The HBIE (2.10), however, has only been collocated at regular points on the boundary of a body.*

2.2 Iterated HBIE and Error Estimation

The heuristic idea that is at the heart of the pointwise error estimation procedure described below (see also [122, 123, 96]) is simple : *the amount by which an approximate solution to the BIE fails to satisfy the HBIE is a measure of the error in the approximation.* The main result of this work is that this heuristic idea, when stated formally, leads to a simple characterization of the error. In essence, the method reduces to finding a second approximation to the solution by iterating the first approximation with the HBIE. This idea is first illustrated in the context of two basic cases : the interior Dirichlet and Neumann problems. Mixed boundary conditions are considered thereafter.

2.2.1 Problem 1 : Displacement Boundary Conditions

Solve the Navier-Cauchy equations:

$$(\lambda + \mu)\nabla(\nabla \cdot \mathbf{u}) + \mu\nabla^2\mathbf{u} = \mathbf{0} \quad \text{in } \mathbf{B}$$

subject to the boundary conditions:

$$\mathbf{u} = \mathbf{f} \quad \text{on } \partial\mathbf{B}$$

This problem is analogous to the Dirichlet problem of potential theory. Under suitable restrictions on the domain, it is possible to prove existence and uniqueness of a solution to this BVP [46].

Either the displacement BIE (2.9) or the traction HBIE (2.10) may be used to formulate a method of solution for the unknown traction on the boundary. The displacement BIE leads to a system of singular integral equations of the first kind for the (unknown) traction:

$$\mathcal{U}_{ij}\tau_j = f_i + \mathcal{T}_{ij}f_j =: g_i^1 \quad i = 1, 2 \tag{2.11}$$

while the traction BIE gives rise to equations of the second kind (for the traction):

$$\tau_i - \mathcal{D}_{ij}^{(N)}\tau_j = -\mathcal{S}_{ij}^{(N)}f_j =: g_i^2 \quad i = 1, 2 \tag{2.12}$$

Recall that $\mathcal{U}_{ij}$ is log-singular, $\mathcal{T}_{ij}$ and $\mathcal{D}_{ij}^{(N)}$ are Cauchy singular, and $\mathcal{S}_{ij}^{(N)}$ is hypersingular. As in potential theory, since $\mathcal{U}_{ij}$ has a logarithmic kernel, one again encounters the problem of the transfinite diameter. For instance, one may show that if the domain B is a circle of radius $\exp[(1/2)(3 - 4\nu)]$, then the BIE (2.11) does not admit a unique solution.

2.2.1.1   Error estimate for the primary problem

Using the two BIEs (2.11) and (2.12), one can formulate an error estimation process that is analogous to the Dirichlet problem in potential theory [96].

- *Step 1:* Solve the displacement BIE (2.11) for the traction $\tau_i^{(1)}$:

$$\mathcal{U}_{ij}\tau_j^{(1)} = (\mathcal{I}_{ij} + \mathcal{T}_{ij})f_j \tag{2.13}$$

  where $\mathcal{I}$ is the identity operator and $\mathcal{I}_{ij}f_j = \delta_{ij}f_j = f_i$, with δ_{ij} the components of the Kronecker delta.

- *Step 2:* Use the traction HBIE (2.12) to iterate the traction and obtain a second approximation $\tau_i^{(2)}$:

$$\tau_i^{(2)} = \mathcal{D}_{ij}^{(N)}\tau_j^{(1)} - \mathcal{S}_{ij}^{(N)}f_j \tag{2.14}$$

This approximation, called the HBIE iterate, will be used for error estimation.

Let the error (in traction) in the primary solution and iterate be:

$$e_i^{\tau(1)} = \tau_i^{(1)} - \tau_i \tag{2.15}$$

$$e_i^{\tau(2)} = \tau_i^{(2)} - \tau_i \tag{2.16}$$

respectively. Define the hypersingular residual to be:

$$r_i^{(\tau)} = \tau_i^{(1)} - \tau_i^{(2)} \tag{2.17}$$

One can now show that:

$$
\begin{aligned}
r_i^{(\tau)} &\overset{(2.17)}{=} \tau_i^{(1)} - \tau_i^{(2)} \\
&\overset{(2.14)}{=} \tau_i^{(1)} - (\mathcal{D}_{ij}^{(N)} \tau_j^{(1)} - \mathcal{S}_{ij}^{(N)} f_j) \\
&\overset{(2.15)}{=} \tau_i - (\mathcal{D}_{ij}^{(N)} \tau_j - \mathcal{S}_{ij}^{(N)} f_j) + e_i^{\tau(1)} - \mathcal{D}_{ij}^{(N)} e_j^{\tau(1)} \\
&\overset{(2.10)}{=} (\mathcal{I}_{ij} - \mathcal{D}_{ij}^{(N)}) e_j^{\tau(1)}
\end{aligned}
$$

so that:

$$r_i^{(\tau)} = (\mathcal{I}_{ij} - \mathcal{D}_{ij}^{(N)}) e_j^{\tau(1)} \tag{2.18}$$

Theorem 1 *For a sufficiently smooth domain, and sufficiently smooth data and solutions (as detailed above), if the solution to the integral equations (2.11) and (2.12) is unique, then there exist real positive constants C_1 and C_2 such that:*

$$C_1 \|r_i^{(\tau)}\| \leq \|e_i^{\tau(1)}\| \leq C_2 \|r_i^{(\tau)}\|$$

Proof: The continuity of the operators is a manifestation of the elliptic nature of the partial differential equation (PDE). Uniqueness of solutions to the integral formulations implies that the operators $(\mathcal{I}_{ij} - \mathcal{D}_{ij}^{(N)})$ and $\mathcal{U}_{ij}$ have continuous inverses [172]. Now use Lemma 1.⋄

2.2.1.2 Error estimate for the iterate

In a manner similar to the previous subsection, one can show that:

$$
\begin{aligned}
e_i^{\tau(2)} &\overset{(2.16)}{=} \tau_i^{(2)} - \tau_i \\
&\overset{(2.14)}{=} \mathcal{D}_{ij}^{(N)} \tau_j^{(1)} - \mathcal{S}_{ij}^{(N)} f_j - \tau_i \\
&\overset{(2.15)}{=} \mathcal{D}_{ij}^{(N)} (e_j^{\tau(1)} + \tau_j) - \mathcal{S}_{ij}^{(N)} f_j - \tau_i \\
&\overset{(2.10)}{=} \mathcal{D}_{ij}^{(N)} e_j^{\tau(1)}
\end{aligned}
$$

so that:

$$e_i^{\tau(2)} = \mathcal{D}_{ij}^{(N)} e_j^{\tau(1)} \tag{2.19}$$

2.2.2   Problem 2 : Traction Boundary Conditions

Solve the Navier-Cauchy equations:

$$(\lambda + \mu)\nabla(\nabla \cdot \mathbf{u}) + \mu\nabla^2\mathbf{u} = \mathbf{0} \quad \text{in } \mathrm{B}$$

subject to the boundary conditions:

$$\mathbf{t} = \mathbf{g} \quad \text{on } \partial\mathrm{B}$$

where the tractions satisfy the consistency conditions of static equilibrium:

$$\int_{\partial B} \mathbf{t}\, ds = \mathbf{0}$$

$$\int_{\partial B} (\mathbf{r} \times \mathbf{t})\, ds = \mathbf{0}$$

It is known that the solution to the above problem exists, and is unique up to a rigid body motion (Fung [46]). The space of two-dimensional rigid body motions may be characterized as (Chen and Zhou [29]):

$$\mathcal{R} := \mathbf{r_0} + \boldsymbol{\omega} \times \mathbf{r} \tag{2.20}$$

where $\mathbf{r_0} \in \mathbf{R}^2$ is a translation, and $\boldsymbol{\omega} = \omega\,\mathbf{k}$ is an axial vector representing a rotation.

The first integral equation formulation for the problem follows from the displacement BIE (2.9). One has an integral equation of the second kind for the (unknown) displacement:

$$u_i + \mathcal{T}_{ij}u_j = \mathcal{U}_{ij}g_j =: h_i^1 \tag{2.21}$$

and using the traction HBIE (2.10):

$$-\mathcal{S}_{ij}^{(N)}u_j = g_i - \mathcal{D}_{ij}^{(N)}g_j =: h_i^2 \tag{2.22}$$

one obtains an integral equation of the first kind for the displacement.

It is important to mention again that the solution of the traction prescribed BVP is arbitrary within a rigid body motion, and, to eliminate this arbitrariness, one must work in a restricted function space as has been done before [96] for Neumann problems in potential theory. An elegant practical way to solve traction prescribed problems in linear elasticity is outlined in a recent paper by Lutz et al. [90] where the singular matrix from the BIE is suitably regularized at the *discretized level* by eliminating rigid body modes.

2.2.2.1 Error estimate for the primary problem

The error estimation technique is analogous to the Neumann problem investigated previously by Menon et al. [96]. First, construct an approximation to the displacement field, $u_i^{(1)}$. Next find $\tau_i^{(2)}$, an iterated approximation to the traction, and use it to estimate the error in the primary solution.

- *Step 1:* Solve the displacement BIE (2.21) for the displacement $u_i^{(1)}$:

$$(\mathcal{I}_{ij} + \mathcal{T}_{ij})u_j^{(1)} = \mathcal{U}_{ij}g_j \tag{2.23}$$

- *Step 2:* Use the traction HBIE (2.10) to obtain $\tau_i^{(2)}$:

$$\tau_i^{(2)} = \mathcal{D}_{ij}^{(N)}g_j - \mathcal{S}_{ij}^{(N)}u_j^{(1)} \tag{2.24}$$

Define the hypersingular residual:

$$r_i^{(\tau)} = \tau_i^{(1)} - \tau_i^{(2)} = g_i - \tau_i^{(2)} \tag{2.25}$$

Also, the error in the displacement is defined as:

$$e_i^{u(1)} = u_i^{(1)} - u_i \tag{2.26}$$

One can now show that:

$$
\begin{aligned}
r_i^{(\tau)} &\overset{(2.25)}{=} g_i - \tau_i^{(2)} \\
&\overset{(2.24)}{=} g_i - (\mathcal{D}_{ij}^{(N)}g_j - \mathcal{S}_{ij}^{(N)}u_j^{(1)}) \\
&\overset{(2.26)}{=} g_i - \mathcal{D}_{ij}^{(N)}g_j + \mathcal{S}_{ij}^{(N)}(u_j + e_j^{u(1)}) \\
&\overset{(2.10)}{=} \mathcal{S}_{ij}^{(N)}e_j^{u(1)}
\end{aligned}
$$

so that:

$$r_i^{(\tau)} = \mathcal{S}_{ij}^{(N)}e_j^{u(1)} \tag{2.27}$$

Theorem 2 *The hypersingular traction residual bounds the error in the displacement globally:*

$$C_1\|r_i^{(\tau)}\| \leq \|e_i^{u(1)}\| \leq C_2\|r_i^{(\tau)}\|$$

Proof: The proof is quite analogous to that of Theorem 1. It follows from using equation (2.27). ⋄

2.2.2.2   The displacement residual

In the traction boundary condition problem, the unknown is the displacement but equation (2.27) relates the traction residual to the error in the displacement. It is proved below, however, that the traction residual is also equal to a suitably defined displacement residual for this problem.

The HBIE (2.10), with u_i added to both sides of it, and upon rearrangement, becomes:

$$u_i = (\mathcal{I}_{ij} - \mathcal{S}_{ij}^{(N)})u_j - (\mathcal{I}_{ij} - \mathcal{D}_{ij}^{(N)})g_j \tag{2.28}$$

Let $u_i^{(1)}$ be the solution of the BIE (2.9). Iterate (2.28) with this solution and define:

$$u_i^{(2)} = (\mathcal{I}_{ij} - \mathcal{S}_{ij}^{(N)})u_j^{(1)} - (\mathcal{I}_{ij} - \mathcal{D}_{ij}^{(N)})g_j \tag{2.29}$$

Define the displacement residual:

$$r_i^{(u)} \equiv u_i^{(1)} - u_i^{(2)} \tag{2.30}$$

One can now show that:

$$
\begin{aligned}
r_i^{(u)} &\equiv u_i^{(1)} - u_i^{(2)} \\
&\overset{(2.29)}{=} u_i^{(1)} - (\mathcal{I}_{ij} - \mathcal{S}_{ij}^{(N)})u_j^{(1)} + (\mathcal{I}_{ij} - \mathcal{D}_{ij}^{(N)})g_j \\
&= \mathcal{S}_{ij}^{(N)}u_j^{(1)} + g_i - \mathcal{D}_{ij}^{(N)}g_j \\
&\overset{(2.10)}{=} \mathcal{S}_{ij}^{(N)}u_j^{(1)} - \mathcal{S}_{ij}^{(N)}u_j \\
&\overset{(2.26)}{=} \mathcal{S}_{ij}^{N}e_j^{u(1)} \\
&\overset{(2.27)}{=} r_i^{(\tau)}
\end{aligned}
\tag{2.31}
$$

so that:

$$r_i^{(u)} = r_i^{(\tau)} \tag{2.32}$$

and $r_i^{(\tau)}$ can be replaced by $r_i^{(u)}$ in Theorem 2 !

Remark 2 *An analogous result in potential theory appears in Menon et al.* *[96]*

2.2.3   Problem 3 : Mixed Boundary Conditions

The general boundary value problem in linear elasticity is:

Solve the Navier-Cauchy equations:

$$(\lambda + \mu)\nabla(\nabla \cdot \mathbf{u}) + \mu\nabla^2\mathbf{u} = \mathbf{0} \quad \text{in } B$$

subject to the boundary conditions:

$$\mathbf{A}\mathbf{u} + \mathbf{B}\mathbf{t} = \mathbf{f} \quad \text{on } \partial B$$

where the matrices $\mathbf{A}$ and $\mathbf{B}$ and the vector $\mathbf{f}$ are prescribed quantities.

This class of problems is the most commonly encountered one in linear elasticity. Indeed, the numerical examples presented in Section 2.4 of this chapter all have mixed boundary conditions imposed upon them. In this case, however, a heuristic approach to error estimation is adopted here.

2.2.3.1 Traction residual

One computes the traction components $\tau_j^{(1)}$ on ∂B by solving the primary BIE (2.9) and then obtains the HBIE iterate $\tau_j^{(2)}$ from the HBIE (2.10). As before, the traction residual is defined as:

$$r_i^{(\tau)} = \tau_i^{(1)} - \tau_i^{(2)} \tag{2.33}$$

The corresponding pointwise error measure is as follows. At a fixed boundary point, if the traction is specified in one direction, and the displacement in the other, then the error in the boundary data is the error in displacement in the first direction, and the error in traction in the second direction. This issue is discussed further in Section 2.3 of this chapter.

2.2.3.2 Stress residual

The stress residual is another important quantity in this work. The primary BIE (2.9) is solved first. This yields the boundary tractions and displacements $\tau_j^{(1)}$ and $u_j^{(1)}$. The boundary stresses $\sigma_{ij}^{(1)}$ are next obtained from the boundary values of the tractions and the tangential derivatives of the displacements, together with Hooke's law. This is a well-known procedure in the BIE literature (see, for example, Mukherjee [98] or Sladek and Sladek [151]).

Next, the iterated boundary stress is obtained from the HBIE (1.36) as follows:

$$\sigma_{ij}^{(2)} = \mathcal{D}_{ijk}\tau_k^{(1)} - \mathcal{S}_{ijk}u_k^{(1)} \tag{2.34}$$

where the required operators are defined in equations (2.5) and (2.6) and the LTB of the above operators are used in equation (2.34). Also, equation (2.34) is collocated only at regular boundary points (where the boundary is locally smooth) inside boundary elements.

One now gets the error in stress, for the BIE and the HBIE iterate, respectively, as:

$$e_{ij}^{s(1)} = \sigma_{ij}^{(1)} - \sigma_{ij} \tag{2.35}$$

$$e_{ij}^{s(2)} = \sigma_{ij}^{(2)} - \sigma_{ij} \tag{2.36}$$

and the stress residual is defined as:

$$r_{ij}^{(s)} = \sigma_{ij}^{(1)} - \sigma_{ij}^{(2)} \tag{2.37}$$

Remark 3 *The stress residual, defined above in equation (2.37), can also be used for problems with displacement or traction boundary conditions, which are special cases of problems with mixed boundary conditions.*

2.3   Element-Based Error Indicators

The main objective of error estimation is the development of suitable element error indicators, which are denoted by η_i. These indicators should satisfy the following criteria:

$$C_1 \eta_i \; \leq \|e\|_{A(\partial B_i)} \; \leq C_2 \eta_i \tag{2.38}$$

$$D_1 \sum_{i=1}^{N} \eta_i^2 \; \leq \|e\|_A^2 \; \leq D_2 \sum_{i=1}^{N} \eta_i^2 \tag{2.39}$$

where A is a suitable norm, $A(\partial B_i)$ denotes the restriction of this norm to the ith element, and C_1, C_2, D_1 and D_2 are appropriate constants. It is often difficult to prove these properties analytically, and one usually takes recourse to numerical experiments. As in the potential theory case [96], this method leads to two natural error indicators. The first is based on the traction residual defined in equation (2.17); the second is based on the stress residual defined in equation (2.37).

These pointwise error measures may be used to define element error indicators. The following are proposed : the first based on the traction residual and the second on the stress residual:

$$\eta_j^{(\tau)} := \|r_i^{(\tau)}\|_{L^2(\partial B_j)} \tag{2.40}$$

$$\eta_j^{(s)} := \|r_{k\ell}^{(s)}\|_{L^2(\partial B_j)} \tag{2.41}$$

Note that the subscript j refers to the jth element - the error indicator is a scalar, not a vector. The L^2 norm is used for convenience and other norms can be used, if desired.

The error estimates, as defined above, do not depend directly on the boundary conditions on an element. The traction residual has been shown to be related to the pointwise error in the boundary unknowns for Dirichlet and Neumann problems in elasticity (Section 2.2, Theorems 1 and 2). Note that, even though the traction residual uses the difference in the primary and iterated tractions, but no explicit information about the displacement, it has been proved in Section 2.2 that the traction residual is equal to the displacement residual for a traction prescribed boundary value problem.

In general, at a local level, on a particular element, the actual error will depend on the boundary conditions. In mixed boundary value problems, for instance, the traction may be prescribed in the x_1-direction and the displacement in the x_2-direction at a boundary point. The errors are, therefore, in the displacement in the x_1-direction, and in the traction in the x_2-direction. Ideally, the traction residual-based element error indicator will capture the L^2 norm of these errors on an element, even for mixed boundary value problems.

The stress residual is also used as a measure of the error in stress on the boundary. At any boundary point in a 2-D problem, at most two components of the stress are known from the prescribed boundary conditions. Thus, there is always some error in a computed stress tensor at a boundary point. Numerical experiments presented below (Section 2.4) suggest that this error is effectively tracked by the stress residual-based error indicator.

2.4   Numerical Examples

Two basic problems from the theory of planar elasticity are considered in this section. The numerical implementation consists of two modules: a standard code for two-dimensional elastostatics, and a set of routines that calculate the hypersingular residual for error estimation. For the first part (i.e. the BIE), a code due to Becker [9] is employed. This code uses collocation with quadratic isoparametric elements. Numerical integration is done using Gaussian quadrature, except on elements that contain the collocation point. Singular integration is avoided using the rigid-body mode, i.e., diagonal terms are evaluated by summing the off-diagonal terms. For the second part of the code (i.e. the HBIE), collocation is carried out at points on the boundary where it is locally smooth, and which are inside boundary elements, in order to determine the components of stress using the traction HBIE. The numerical method used for evaluation of the necessary hypersingular integrals here is due to Guiggiani [60]. In the following examples, the stress tensor at a boundary point is evaluated by using the HBIE (1.36) at three boundary points inside a boundary element, and then a quadratic polynomial is employed to approximate the stress components over each element.

2.4.1   Example 1: Lamé's Problem of a Thick-Walled Cylinder under Internal Pressure

Consider an infinitely long hollow cylinder subjected to an internal pressure $p = 1$. The inner radius $r_i = 3$, and the outer radius $r_o = 6$. Material properties are also chosen of $O(1)$, namely Young's modulus, $E = 1.0$, and Poisson's ratio $\nu = 0.3$. Consistent units are assumed throughout this paper.

Symmetry is employed and the problem is formulated as a mixed boundary value problem on a quarter of the cylinder. The mesh used to solve this problem is shown in Figure 2.1(a). Notice that the mesh is not biased *a priori* in the sense that the element density is not increased on parts of the boundary where the error is expected to be high.

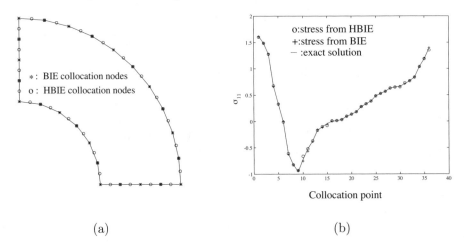

(a)   (b)

Figure 2.1: (a) Discretized domain for the Lamé problem with 12 elements. (b) Comparison of analytical and numerical solutions for σ_{11} (from [127])

Figure 2.1(b) presents a comparison of the computed and analytical solutions for the stress component σ_{11} at the collocation points used for the HBIE. Note that the continuous line for the exact solution is just used as a matter of convenience and does not have any meaning except at discrete points, since the x-axis is the collocation point number. The results for σ_{12} and σ_{22} display similar accuracy and are not shown here.

More importantly, a pointwise comparison between the absolute value of the error and pointwise measurements of the hypersingular residual is considered next. Unlike the numerical examples in potential theory [96], the hypersingular residual and error are often of opposite signs in this elasticity example. Also, it is seen that the residual is not an upper bound as it underestimates the error at some points. Since the stress residual is a symmetric tensor with three independent components, a comparison between the pointwise error and stress residual in each direction is carried out here. The error in stress, for the BIE and the HBIE iterate, and the stress residual, are defined in (2.35 - 2.37).

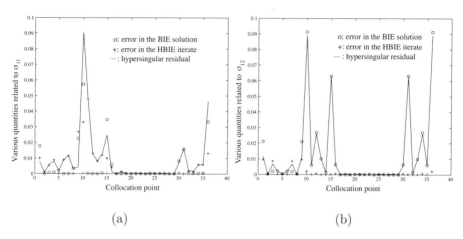

(a)   (b)

Figure 2.2: Absolute values of error in the BIE solution, error in the HBIE iterate, and the hypersingular residual, for (a) σ_{11} and (b) σ_{12}. All variables are unscaled (from [127])

Now consider a comparison of absolute values of errors, and the residual in σ_{11} in Figure 2.2(a), and in σ_{12} in Figure 2.2(b). It is seen that the stress residual provides good pointwise tracking of the error on a relatively coarse mesh.

Of most practical importance (e.g. in adaptivity) is the performance of element error indicators. In particular, the performance of the two indicators $\eta_j^{(\tau)}$ and $\eta_j^{(s)}$ defined in equations (2.40) and (2.41), respectively, is studied here. The first is a traction residual-based error indicator, and the second uses the stress residual.

The element error indicator based on the traction residual ($\eta_j^{(\tau)}$ from equation (2.40)) is compared with the element-based L^2 norm of the error in the unspecified boundary data in Figure 2.3(a). On the other hand, the element error indicator based on the stress residual ($\eta_j^{(s)}$ from equation (2.41)) is compared to the element-based L^2 norm of the error in stress, on all the boundary elements, in Figure 2.3(b). The stress residuals are seen to capture the error trends quite effectively.

Remark 4 *The comparison between the traction residual and error in the displacement is difficult unless one uses normalized values. A simple way to do this is to use nondimensional quantities to begin with. This is the approach followed in this work.*

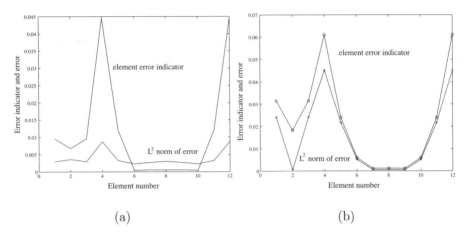

(a)   (b)

Figure 2.3: Comparisons of element error indicators with actual errors, for (a) traction residual-based error with error in the unspecified boundary data, and (b) stress residual-based error with error in boundary stresses (from [127])

2.4.2   Example 2: Kirsch's Problem of an Infinite Plate with a Circular Cutout

Consider an infinite plate with a circular cutout of radius 1, subject to a traction $\tau_1 = 1$ at infinity. Material properties are the same as in the previous example.

The displacement and stress fields for this problem may be found in Timoshenko and Goodier [167]. In order to simulate this problem with a finite geometry, the boundaries of a finite square domain are subjected to tractions computed from the exact solution of an infinite plate subjected to traction at infinity. Using symmetry, only a quarter of the plate is used in the computer model. The mesh is shown in Figure 2.4.

Pointwise comparisons between the errors (in the BIE solution and in the HBIE iterate), and the residual, in some components of the stress, are presented first. Figures 2.5(a) and 2.5(b) show the absolute values of these errors in two stress components, together with the corresponding stress residuals. Of course, the error in BIE stress component is zero at points where that particular stress component is prescribed as a boundary condition.

The more important comparison is between the computed element error indicators and the L^2 norms of the error on each element (Figure 2.6). Figure 2.6(a) uses errors in the unspecified boundary data while Figure 2.6(b) uses errors in stress components. It is seen that the traction-based error indicator underestimates the error on some elements. The stress-based error indicator performs better and accurately captures the error in stress on most of the elements.

In conclusion, the error estimation method presented here has the advantage of capturing errors in the stress field. On the other hand, the usual residual

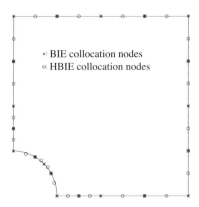

Figure 2.4: Discretized domain for the problem of a plate with a cutout with
10 elements. Plate side is 4 units and cutout radius is 1 unit of length (from
[127])

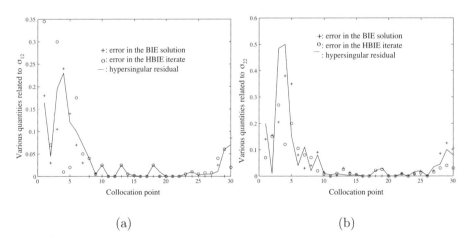

(a)   (b)

Figure 2.5: Absolute values of error in the BIE solution, error in the HBIE
iterate, and the hypersingular residual for (a) σ_{12} and (b) σ_{22}. All variables are
unscaled (from [127])

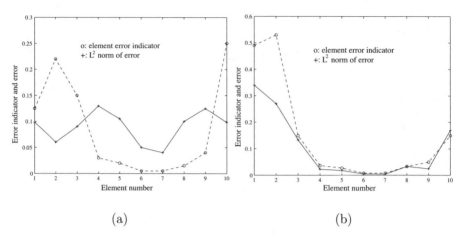

(a)   (b)

Figure 2.6: Comparisons of element error indicators with actual errors, for (a) traction residual-based error with error in the unspecified boundary data, and (b) stress residual-based error with error in boundary stresses (from [127])

techniques may only be used to compute the error in displacements. On physical grounds, a measure of the error in stress is preferable to a measure of displacement error. It may also be viewed as a stronger measure of convergence - i.e. the approximate displacement field, *and its gradient* have converged to the actual solution in this case.

Chapter 3

THIN FEATURES

Many boundary value problems that are solved by the BEM involve thin features. Common examples are cracks and thin plates and shells. This chapter first presents BEM formulations for 3-D potential theory in a region exterior to thin plates, for applications in micro-electro-mechanical systems (MEMS). This is followed by a discussion of crack problems (linear elastic fracture mechanics - LEFM) in 3-D linear elasticity.

3.1 Exterior BIE for Potential Theory: MEMS

Exterior BIEs for potential theory, suitable for applications in MEMS, are presented in this section. Numerical results for a simple system with two thin conducting plates follow. Further details are available in [6].

3.1.1 Introduction to MEMS

The field of micro-electro-mechanical systems (MEMS) is a very broad one that includes fixed or moving microstructures; encompassing micro-electro-mechanical, microfluidic, micro-opto-electro-mechanical and micro-thermal-mechanical devices and systems. MEMS usually consists of released microstructures that are suspended and anchored, or captured by a hub-cap structure and set into motion by mechanical, electrical, thermal, acoustical or photonic energy source(s).

Typical MEMS structures consist of arrays of thin beams with cross-sections in the order of microns (μm) and lengths in the order of ten to hundreds of microns. Sometimes, MEMS structural elements are plates. An example is a small rectangular silicon plate with sides in the order of mm and thickness of the order of microns, that deforms when subjected to electric fields. Owing to its small size, significant forces and/or deformations can be obtained with the application of low voltages ($\approx$ 10 volts). Examples of devices that utilize vibrations of such plates are synthetic microjets ([142, 8] - for mixing, cooling of electronic components, micropropulsion and flow control), microspeakers [71] etc.

Numerical simulation of electrically actuated MEMS devices have been carried out for around a decade or so by using the BEM to model the exterior electric field and the FEM (see, e.g. [179, 190, 66]) to model deformation of the structure. The commercial software package MEMCAD [147], for example, uses the commercial FEM software package ABAQUS for mechanical analysis, together with a BEM code FastCap [112] for the electric field analysis. Other examples of such work are [49, 148, 1]; as well as [147, 149] for dynamic analysis of MEMS.

The focus of this section is the BEM analysis of the electric field exterior to very thin conducting plates. A convenient way to model such a problem is to assume plates with vanishing thickness and solve for the sum of the charges on the upper and lower surfaces of each plate [61]. The standard BIE with a weakly singular kernel is used here and this approach works well for determining, for example, the capacitance of a parallel plate capacitor. For MEMS calculations, however, one must obtain the charge densities separately on the upper and lower surfaces of a plate since the traction at a surface point on a plate depends on the square of the charge density at that point. The gradient BIE is employed in the present work to obtain these charge densities separately. *Careful regularization of the gradient equation, to take care of singular and nearly singular integrals that arise, is the principal contribution of the present work.* The work of Liu [87], on thin shells, is of great value to the research reported here. Gray [53] and Nishimura and his coworkers [118, 180] have considered the 3-D Laplace equation in a region exterior to a narrow slit or crack. Gray addresses applications in electroplating problems. The research described in Refs. [118, 180] is quite different from the problem under consideration here. Their primary interest is in the crack opening displacement with zero normal displacement derivative on the crack faces, while the separate charge densities (that are proportional to the normal derivative of the potential) are of interest in this chapter. The formulation given in the present work is a BEM scheme that is particularly well suited for MEMS analysis of very thin plates - for $h/L \leq .001$ - in terms of the length L (of a side of a square plate) and its thickness h. A similar approach has also been developed for MEMS and nano-electro-mechanical systems (NEMS) with very thin beams [7], but this work is not presented in this book.

As a byproduct of the development of the thin plate BEM, an enhanced BEM, suitable for MEMS analysis of moderately thick plates, has also been developed in this work. It is shown that accurate evaluation of weakly singular and nearly weakly singular integrals plays a key role here.

This section is organized as follows. The usual and gradient BIEs for potential theory, in an infinite region exterior to a structure composed of thin conducting plates, are first presented and regularized. Singular and nearly singular integrals, both weak and strong, are discussed next. Numerical results are presented and discussed for a model problem (a parallel plate capacitor) from three methods - the usual BEM, the enhanced BEM and the thin plate BEM.

3.1.2 Electric Field BIEs in a Simply Connected Body

First consider the solution of Laplace's equation in a three-dimensional (3-D) simply connected body.

3.1.2.1 Usual BIE - indirect formulation

Referring to Figure 1.1, for a source point $\boldsymbol{\xi} \in B$ (with bounding surface ∂B), one has the usual indirect BIE:

$$\phi(\boldsymbol{\xi}) = \int_{\partial B} \frac{\nu(\mathbf{y})}{4\pi r(\boldsymbol{\xi}, \mathbf{y})\epsilon} ds(\mathbf{y}) \tag{3.1}$$

where ϕ is the potential, $\mathbf{r}(\boldsymbol{\xi}, \mathbf{y}) = \mathbf{y} - \boldsymbol{\xi}$, $r = |\mathbf{r}|$, ϵ is the dielectric constant of the medium and ν is the (unknown) surface density function on ∂B.

3.1.2.2 Gradient BIE - indirect formulation

Taking the gradient of ϕ at the source point $\boldsymbol{\xi}$ results in:

$$\boldsymbol{\nabla}_{\boldsymbol{\xi}}\phi(\boldsymbol{\xi}) = \int_{\partial B} \frac{\nu(\mathbf{y})}{4\pi\epsilon} \boldsymbol{\nabla}_{\boldsymbol{\xi}} \left(\frac{1}{r(\boldsymbol{\xi}, \mathbf{y})} \right) ds(\mathbf{y}) = \int_{\partial B} \frac{\nu(\mathbf{y})\mathbf{r}(\boldsymbol{\xi}, \mathbf{y})}{4\pi r^3(\boldsymbol{\xi}, \mathbf{y})\epsilon} ds(\mathbf{y}) \tag{3.2}$$

Alternatively, one can write (3.2) as:

$$\frac{\partial \phi}{\partial \xi_k}(\boldsymbol{\xi}) = \int_{\partial B} \frac{\nu(\mathbf{y})(y_k - \xi_k)}{4\pi r^3(\boldsymbol{\xi}, \mathbf{y})\epsilon} ds(\mathbf{y}) \tag{3.3}$$

Note that, in general, the function $\nu(\mathbf{y})$ is not the charge density. It becomes equal to the charge density when B is the infinite region exterior to the conductors. This is discussed in Section 3.1.3.

3.1.3 BIES in Infinite Region Containing Two Thin Conducting Plates

Now consider the situation shown in Figure 3.1. Of interest is the solution of the following Dirichlet problem for Laplace's equation:

$$\nabla^2\phi(\mathbf{x}) = 0, \quad \mathbf{x} \in B, \quad \phi(\mathbf{x}) \text{ prescribed for } \mathbf{x} \in \partial B \tag{3.4}$$

where B is now the region *exterior* to the two plates. The unit normal $\mathbf{n}$ to ∂B is defined to point away from B (i.e. into a plate).

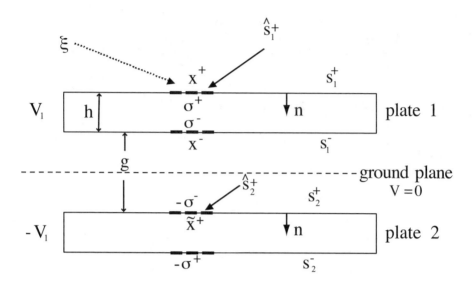

Figure 3.1: Parallel plate capacitor with two plates (from [6])

3.1.3.1   Regular BIE - source point approaching a plate surface S_1^+

Let $\hat{S}_1^+ \subset S_1^+$ be a small neighborhood of $\mathbf{x}^+$. As $\boldsymbol{\xi} \to \mathbf{x}^+ \in \hat{S}_1^+ \subset S_1^+$ (see Figure 3.1), one has:

$$
\begin{aligned}
\phi(\mathbf{x}^+) &= \int_{S_1^+ - \hat{S}_1^+} \frac{\beta(\mathbf{y})}{4\pi r(\mathbf{x}^+, \mathbf{y})\epsilon} dS(\mathbf{y}) + \int_{\hat{S}_1^+} \frac{\beta(\mathbf{y})}{4\pi r(\mathbf{x}^+, \mathbf{y})\epsilon} dS(\mathbf{y}) \\
&\quad + \int_{S_2^+} \frac{\beta(\mathbf{y})}{4\pi r(\mathbf{x}^+, \mathbf{y})\epsilon} dS(\mathbf{y})
\end{aligned} \tag{3.5}
$$

Here $\beta(\mathbf{y}) = \sigma(\mathbf{y}^+) + \sigma(\mathbf{y}^-)$, where σ is now the charge density at a point on a plate surface. The second integral above is weakly singular, while the rest are usually regular. It should be noted, however, that the last integral above becomes nearly weakly singular when both h and g are small.

A similar equation can be written for $\mathbf{x}^+ \in S_2^+$. For the case shown in Figure 3.1, however, this is not necessary since $\beta(\mathbf{y})$ is equal and opposite on the two plates. Therefore, for this case, equation (3.5) is sufficient to solve for β on both the plates !

3.1.3.2   Gradient BIE - source point approaching a plate surface S_1^+

The governing equation.   It is first noted that for $\mathbf{x}^+ \in S_k^+ \cup S_k^-$, $k = 1, 2$:

$$
\sigma(\mathbf{x}) = \epsilon \frac{\partial \phi}{\partial n}(\mathbf{x}) = \epsilon \mathbf{n}(\mathbf{x}) \cdot [\boldsymbol{\nabla}_{\boldsymbol{\xi}} \phi(\boldsymbol{\xi})]_{\boldsymbol{\xi} = \mathbf{x}} \tag{3.6}
$$

Consider the limit $\boldsymbol{\xi} \to \mathbf{x}^+ \in \hat{S}_1^+ \subset S_1^+$. It is important to realize that this limit is meaningless for a point $\mathbf{x}$ on the edge of a plate, since the charge density is singular on its edges. One has:

$$
\begin{aligned}
\sigma(\mathbf{x}^+) &= \int_{S_1^+ - \hat{S}_1^+} \frac{\beta(\mathbf{y})\mathbf{r}(\mathbf{x}^+,\mathbf{y}) \cdot \mathbf{n}(\mathbf{x}^+)}{4\pi r^3(\mathbf{x}^+,\mathbf{y})} dS(\mathbf{y}) \\
&+ \int_{\hat{S}_1^+} \frac{\mathbf{r}(\mathbf{x}^+,\mathbf{y}) \cdot [\beta(\mathbf{y})\mathbf{n}(\mathbf{x}^+) - \beta(\mathbf{x})\mathbf{n}(\mathbf{y})]}{4\pi r^3(\mathbf{x}^+,\mathbf{y})} dS(\mathbf{y}) \\
&+ \frac{\beta(\mathbf{x})}{4\pi}\Omega(\hat{S}_1^+,\mathbf{x}^+) + \int_{S_2^+} \frac{\beta(\mathbf{y})\mathbf{r}(\mathbf{x}^+,\mathbf{y}) \cdot \mathbf{n}(\mathbf{x})}{4\pi r^3(\mathbf{x}^+,\mathbf{y})} dS(\mathbf{y}) \quad (3.7)
\end{aligned}
$$

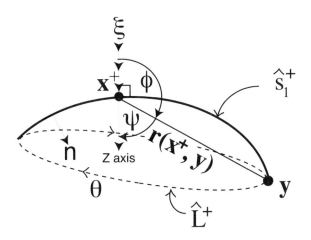

Figure 3.2: Line integral for evaluation of solid angle (from [6])

In the above, the solid angle subtended by the surface element $\hat{S}_1^+$ at the point $\mathbf{x}^+$ is (see [87, 105] and Figure 3.2):

$$
\Omega(\hat{S}_1^+,\mathbf{x}^+) = \fint_{\hat{S}_1^+} \frac{\mathbf{r}(\mathbf{x}^+,\mathbf{y}) \cdot \mathbf{n}(\mathbf{y})}{r^3(\mathbf{x}^+,\mathbf{y})} dS(\mathbf{y}) = \int_0^{2\pi} [1 - \cos(\psi(\theta))]d\theta \quad (3.8)
$$

where the symbol $\fint$ denotes the Finite Part (FP) of the integral in the sense of Mukherjee [101, 102].

Equations (3.7) and (3.8) give the final equation:

$$
\frac{1}{2}[\sigma(\mathbf{x}^+) - \sigma(\mathbf{x}^-)] = \int_{S_1^+ - \hat{S}_1^+} \frac{\beta(\mathbf{y})\mathbf{r}(\mathbf{x}^+,\mathbf{y}) \cdot \mathbf{n}(\mathbf{x}^+)}{4\pi r^3(\mathbf{x}^+,\mathbf{y})} dS(\mathbf{y})
$$

$$+ \int_{\hat{S}_1^+} \frac{\mathbf{r}(\mathbf{x}^+, \mathbf{y}) \cdot [\beta(\mathbf{y})\mathbf{n}(\mathbf{x}^+) - \beta(\mathbf{x})\mathbf{n}(\mathbf{y})]}{4\pi r^3(\mathbf{x}^+, \mathbf{y})} dS(\mathbf{y})$$

$$- \frac{\beta(\mathbf{x})}{4\pi} \int_0^{2\pi} \cos(\psi(\theta)) d\theta + \int_{S_2^+} \frac{\beta(\mathbf{y})\mathbf{r}(\mathbf{x}^+, \mathbf{y}) \cdot \mathbf{n}(\mathbf{x}^+)}{4\pi r^3(\mathbf{x}^+, \mathbf{y})} dS(\mathbf{y}) \quad (3.9)$$

Here (see Figure 3.2), a local coordinate system (x, y, z) is set up with the origin at $\mathbf{x}^+$ such that the positive z axis intersects the surface $\hat{S}_1^+$. Now, ψ is the angle between the positive z axis and $\mathbf{r}(\mathbf{x}^+, \mathbf{y})$ with $\mathbf{y} \in \hat{L}^+$, and θ the angle between the positive x axis and the projection of $\mathbf{r}(\mathbf{x}^+, \mathbf{y})$ in the xy plane.

In the above, the second integral on the right-hand side is weakly singular, while the rest are usually regular. The last integral above, however, becomes nearly strongly singular if both the thickness h and the gap g are small. Once β is known on both plates, (3.9) can be used, *as a postprocessing step*, to obtain σ^+ and σ^- on both plates.

Equation (3.9) for flat plates. First, it should be emphasized that equations (3.5) and (3.9) are valid for thin curved shells as well as for flat plates. For a pair of symmetric flat plates (see Figure 3.1), the first, second and third integrals on the right-hand side of (3.9) vanish, and one is left with the simple equation:

$$\tfrac{1}{2}[\sigma(\mathbf{x}^+) - \sigma(\mathbf{x}^-)] = \int_{S_2^+} \frac{\beta(\mathbf{y})\mathbf{r}(\mathbf{x}^+, \mathbf{y}) \cdot \mathbf{n}(\mathbf{x}^+)}{4\pi r^3(\mathbf{x}^+, \mathbf{y})} dS(\mathbf{y}) \quad (3.10)$$

Equation (3.10) implies that $\sigma(\mathbf{x}^+) = \sigma(\mathbf{x}^-)$ if one has only one plate. This, of course, is true. The existence of the second plate in Figure 3.1 is the reason for (in general) $\sigma(\mathbf{x}^+) \neq \sigma(\mathbf{x}^-)$.

3.1.3.3   Two plates very close together

For cases in which the gap g between the thin plates in Figure 3.1 is also of the order of the (small) plate thickness, the last integral on the right-hand side of equation (3.5) must be treated as nearly weakly singular. In this case, this integral should be written as:

$$\int_{S_2^+} \frac{\beta(\mathbf{y})}{4\pi r(\mathbf{x}^+, \mathbf{y})\epsilon} dS(\mathbf{y}) = \int_{S_2^+ - \hat{S}_2^+} \frac{\beta(\mathbf{y})}{4\pi r(\mathbf{x}^+, \mathbf{y})\epsilon} dS(\mathbf{y})$$

$$+ \int_{\hat{S}_2^+} \frac{\beta(\mathbf{y}) - \beta(\tilde{\mathbf{x}})}{4\pi r(\mathbf{x}^+, \mathbf{y})\epsilon} dS(\mathbf{y}) + \frac{\beta(\tilde{\mathbf{x}})}{4\pi\epsilon} \int_{\hat{S}_2^+} \frac{1}{r(\mathbf{x}^+, \mathbf{y})} dS(\mathbf{y}) \quad (3.11)$$

where $\tilde{\mathbf{x}}^+ \in \hat{S}_2^+$, $\tilde{\mathbf{x}}^- \in \hat{S}_2^-$ and $\beta(\tilde{\mathbf{x}}) = \sigma(\tilde{\mathbf{x}}^+) + \sigma(\tilde{\mathbf{x}}^-)$. The first and second integrals on the right-hand side of (3.11) are regular. (The second integral is $\mathcal{O}(\tilde{r}/r)$ where $\tilde{r} = |\mathbf{y} - \tilde{\mathbf{x}}^+|$. As $\tilde{r} \to 0$, $r \to g + h$, so that this integrand

$\rightarrow 0$.) The last integral is nearly singular. A procedure for accurate evaluation of nearly singular integrals is presented in Section 3.1.4.

Also, the last integral on the right-hand side of (3.9) now becomes nearly strongly singular. This integral, called J, can be evaluated as follows. One can write:

$$
\begin{aligned}
J = & \int_{S_2^+ - \hat{S}_2^+} \frac{\beta(\mathbf{y})\mathbf{r}(\mathbf{x}^+, \mathbf{y}) \cdot \mathbf{n}(\mathbf{x}^+)}{4\pi r^3(\mathbf{x}^+, \mathbf{y})} dS(\mathbf{y}) \\
& + \int_{\hat{S}_2^+} \frac{\mathbf{r}(\mathbf{x}^+, \mathbf{y}) \cdot [\beta(\mathbf{y})\mathbf{n}(\mathbf{x}^+) - \beta(\tilde{\mathbf{x}})\mathbf{n}(\mathbf{y})]}{4\pi r^3(\mathbf{x}^+, \mathbf{y})} dS(\mathbf{y}) \\
& + \frac{\beta(\tilde{\mathbf{x}})}{4\pi} \Omega(\hat{S}_2^+, \mathbf{x}^+)
\end{aligned}
\tag{3.12}
$$

where (see Figure 3.2):

$$
\Omega(\hat{S}_2^+, \mathbf{x}^+) = \int_{\hat{S}_2^+} \frac{\mathbf{r}(\mathbf{x}^+, \mathbf{y}) \cdot \mathbf{n}(\mathbf{y})}{r^3(\mathbf{x}^+, \mathbf{y})} dS(\mathbf{y}) = \int_0^{2\pi} [1 - \cos(\psi(\theta))] d\theta
\tag{3.13}
$$

It is noted that, in this case, the point $\mathbf{x}^+$ is slightly above $\hat{S}_2^+$ and that the second term in (3.13) denotes a "nearly FP" integral.

The idea of regularizing (3.12) with $\beta(\tilde{\mathbf{x}})$ has been inspired by earlier work on evaluation of nearly singular integrals [104].

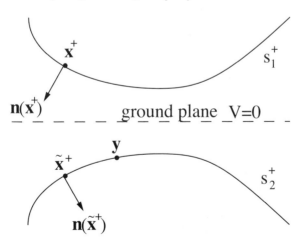

Figure 3.3: Symmetric deformation of two plates (from [6])

Let the integrals on the right-hand side of (3.12) be called J_1, J_2, J_3. Each of the three integrals is regular. The fact that the second integral J_2 is regular can be proved as follows.

It is assumed that the two plates always remain symmetric with respect to the ground plane, even after deformation (see Figure 3.3). Their equations, therefore, are of the form $x_3 = \pm f(x_1, x_2)$. One now has:

$$\mathbf{n}(\mathbf{x}^+) \propto -\mathbf{k} + \mathbf{i}f_{,1} + \mathbf{j}f_{,2}, \qquad \mathbf{n}(\tilde{\mathbf{x}}^+) \propto -\mathbf{k} - \mathbf{i}f_{,1} - \mathbf{j}f_{,2} \qquad (3.14)$$

As $\mathbf{y} \to \tilde{\mathbf{x}}^+$,

$$[\beta(\mathbf{y})\mathbf{n}(\mathbf{x}^+) - \beta(\tilde{\mathbf{x}})\mathbf{n}(\mathbf{y})] \quad \propto$$
$$\mathbf{k}[\beta(\tilde{\mathbf{x}}) - \beta(\mathbf{y})] + [\mathbf{i}a + \mathbf{j}b][\beta(\tilde{\mathbf{x}}) + \beta(\mathbf{y})] \qquad (3.15)$$

where $\mathbf{i}, \mathbf{j}, \mathbf{k}$ are Cartesian unit vectors and a and b are some numbers.

As $\mathbf{y} \to \tilde{\mathbf{x}}^+$, $\mathbf{r}(\mathbf{x}^+, \mathbf{y}) \propto -\mathbf{k}$, so that the integrand of J_2 is $\mathcal{O}(\tilde{r}/r^2)$ where $\tilde{r} = |\mathbf{y} - \tilde{\mathbf{x}}^+|$. In this limit, $\tilde{r} \to 0$, $r \to g + h$, so that the integrand of $J_2 \to 0$.

3.1.3.4   Consistency check

It is interesting to examine the forms of equations (3.5) and (3.9) when collocated at $\mathbf{x}^-$ on S_1^- (see Figure 3.1). Equation (3.5) yields $\phi(\mathbf{x}^+) = \phi(\mathbf{x}^-)$. Referring to equation (3.7), its left-hand side becomes $\sigma(\mathbf{x}^-)$, and the signs of the first, second and fourth terms on its right-hand side change (since $\mathbf{n}(\mathbf{x}^-) = -\mathbf{n}(\mathbf{x}^+)$ and $\mathbf{n}(\mathbf{y})$ for $\mathbf{y} \in S_1^+$ equals $-\mathbf{n}(\mathbf{y})$ for $\mathbf{y} \in S_1^-$). The solid angle expression (see (3.8)) now becomes:

$$\Omega(\hat{S}_1^{\,-}, \mathbf{x}^-) = \int_0^{2\pi} [1 + \cos(\psi(\theta))] d\theta \qquad (3.16)$$

Using these facts, it is easy to show that, as expected, equation (3.9) remains unchanged when collocated at $\mathbf{x}^-$!

3.1.4   Singular and Nearly Singular Integrals

Certain BEM integrals require special care for thin plates and when thin plates come close together. The usual BEM must deal with weakly and nearly weakly singular integrals in such cases, while the thin plate BEM must deal with both nearly weakly singular (the last integral on the right-hand side of (3.5)) and nearly strongly singular (the last integral on the right-hand side of (3.9)) integrals; as well as weakly and strongly singular integrals. The weakly singular case involving this kernel has been addressed before by many authors (see, e.g. [114, 26]). The nearly (also called quasi) weakly singular case, along with other nearly singular integrals of various orders, can be effectively evaluated by employing a cubic polynomial transformation due to Telles [165] and Telles and Oliveira [166]. Several other authors have also considered similar problems (e.g. [64, 104]) - many of these references are available in [165, 114, 166, 104] and are not repeated here in the interest of brevity. A new simple approach for evaluation of nearly weakly singular integrals is presented below.

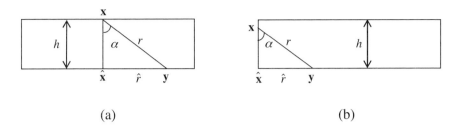

(a)   (b)

Figure 3.4: Nearly weakly singular integrals (from [6])

Proposed new method for evaluating nearly weakly singular integrals.
A typical situation where one encounters weakly and nearly weakly singular
integrals in the usual BEM is shown in Figure 3.4. Consider a source point $\mathbf{x}$ on
the top face of a plate and its image point $\hat{\mathbf{x}}$ on the bottom face in Figure 3.4(a).
Two kinds of singular ($\mathcal{O}(1/r)$) integrals arise - a weakly singular integral on the
boundary element Δ on the top face of the plate that contains $\mathbf{x}$, and, since h is
small, a nearly weakly singular integral on the boundary element $\hat{\Delta}$ (the image
of Δ) on the bottom face of the plate that contains $\hat{\mathbf{x}}$. The weakly singular
integral is evaluated by employing the methods outlined in [114, 26]. A nearly
weakly singular integral (see above and also the last term on the right-hand
side of (3.11)) has the form:

$$I(\mathbf{x}) = \int_{\hat{\Delta}} \frac{\sigma(\mathbf{y})ds(\mathbf{y})}{4\pi\epsilon r(\mathbf{x},\mathbf{y})} \tag{3.17}$$

The integrand above is multiplied by $\hat{r}/\hat{r}$ with the result:

$$I(\mathbf{x}) = \int_{\hat{\Delta}} \frac{[\sigma(\mathbf{y})(\hat{r}/r)]ds(\mathbf{y})}{4\pi\epsilon\hat{r}(\hat{\mathbf{x}},\mathbf{y})} \tag{3.18}$$

Since $\hat{r}/r$ is $\mathcal{O}(1)$ and $\to 0$ as $\mathbf{y} \to \hat{\mathbf{x}}$ (i.e. as $\hat{r} \to 0$), the integrand in (3.18)
is weakly singular, of $\mathcal{O}(1/\hat{r})$ as $\hat{r} \to 0$. Therefore, the integral (3.18) can be
evaluated by employing the methods described in [114, 26].

The source point $\mathbf{x}$ may also lie on a side face of the plate (in the 3-D BEM)
as shown in Figure 3.4(b). The same idea (3.18) can be applied in this case as
well.

Performance of new method for nearly weakly singular integrals.
The performance of the new method is compared with that of standard Gauss
integration. Figure 3.5(a) shows the source point and region of integration (a
triangle). The triangle is purposely chosen to be fairly elongated. Numerical
results (for $\sigma = 1$ and $\epsilon = 1$) appear in Figure 3.5(b). It is seen that for
$h < 1/100$, standard Gauss integration, even with 19 Gauss points, cannot reduce the error below around 4%. The new method is seen to take care of these
nearly weakly singular integrals very well, even for very small values of h.

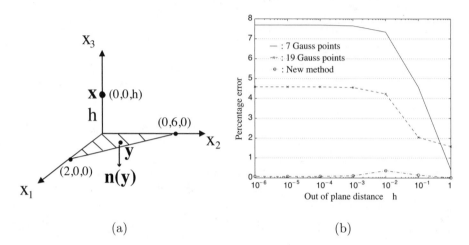

(a)   (b)

Figure 3.5: Nearly weakly singular integral over a triangle. (a) Schematic (b) Errors in numerical results from various methods (from [6])

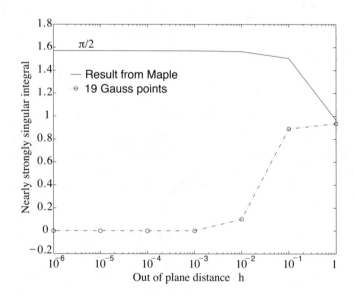

Figure 3.6: Nearly strongly singular integral (from [6])

Nearly strongly singular integrals. When two plates get close together, one encounters nearly strongly singular integrals of the form (see (3.13)):

$$I = \int_\Delta \frac{\mathbf{r}(\mathbf{x}, \mathbf{y}) \cdot \mathbf{n}(\mathbf{y})}{r^3(\mathbf{x}, \mathbf{y})} ds(\mathbf{y}) = \int_\Delta \frac{h}{r^3(\mathbf{x}, \mathbf{y})} ds(\mathbf{y}) \qquad (3.19)$$

where the second expression above is the special version for the situation depicted in Figure 3.5(a). It is absolutely critical that this integral be evaluated carefully. The pathetic result of applying standard Gauss integration (even with 19 Gauss points) appears in Figure 3.6. The correct results in this figure are obtained from the symbolic computational code Maple [65]. A Stoke transformation to convert this nearly strongly singular surface integral to a line integral [87] is recommended in this work (see (3.13)).

3.1.5 Numerical Results

3.1.5.1 Three BEM Models

Three BEM models, the usual, the enhanced and the thin plate, are of concern to this work. These are briefly described below.

Usual BEM. The usual BEM is the standard version with weakly singular integrals evaluated by the method outlined in [26]. For the sake of simplicity and consistency between different BEM models, only uniform BEM meshes, composed of $T6$ triangles, are used on the square faces of objects in this work. It is well known that the charge density is singular on the edges of a plate. Therefore, nonconforming boundary elements should be used and collocation points should not be placed at plate edges. This procedure, however, makes the BEM code somewhat cumbersome and expensive. For simplicity, regular $T6$ elements are used on square plate faces, in all three types of BEM models, with collocation points placed everywhere including on plate edges. Further, in the interest of standardization, the usual BEM only uses $T6$ elements in this work.

The usual BEM code is first verified by solving for the charge distribution on the surface of a unit cube, subjected to unit surface potential. The capacitance (the total charge on the cube surface divided by the voltage) (with $\epsilon = 1$) from the BEM is 8.28 while that from FastCap [113] is 8.3. The BEM mesh for this problem has an 8×8 array of squares on each face, each divided into two $T6$ triangles, for a total of 768 boundary elements.

Enhanced BEM. The enhanced BEM is designed to solve problems with moderately thick plates accurately and efficiently. Like the usual BEM, weakly singular integrals are again evaluated by the methods outlined in [26]. The enhanced BEM has two additional features:

- Nearly weakly singular integrals are evaluated by the method outlined in Section 3.1.4 above.

- Detailed modeling of side faces for small values of h/L is not desirable
  for two reasons. One reason is that this would either lead to triangular
  boundary elements (on these side faces) with very large aspect ratios,
  or, alternatively, to a prohibitively large number of boundary elements.
  The second reason is a matter of efficiency. In the enhanced BEM, $T3$
  elements are used on the side faces. This assures linear interpolation
  of the charge density across a side face with no new unknowns being
  introduced on these faces (and, therefore, no additional nearly weakly
  singular integrals). Obviously, this idea is good for moderate values of
  h/L. It is not a good idea for cubelike conductors. Also, the usual as well
  as the enhanced BEM breaks down for very small values of h/L. This
  issue is discussed further in Section 3.1.6 below.

Thin plate BEM.   This is the model that has been presented in detail in
Section 3.1.3 of this chapter. In this case, nearly singular integrals only arise
when plates are very close together (i.e. for small values of g/L). Again, weakly
singular and nearly weakly singular integrals (when they arise) are evaluated
by the method outlined in [26], and in Section 3.1.4, respectively. Strongly
singular integrals and nearly strongly singular integrals (when they arise) are
evaluated by Stoke regularization (see equations (3.8) and (3.13)). The value
of $g/L = 0.2$ in many examples below. The corresponding integrals are treated
as regular in these cases.

Collocation points.   It is important to point out that, in general, the usual
BEM requires collocation points on the entire bounding surface of a plate, the
enhanced BEM on the upper and lower surfaces, and the thin plate BEM only
on the upper surface. Of course, symmetry can be used, whenever possible, to
further reduce the number of collocation points.

3.1.6   The Model Problem - a Parallel Plate Capacitor

Numerical results are obtained from three methods - the usual, enhanced and
thin plate BEM. The model problem is a parallel plate capacitor with two
square plates (Figure 3.1). The length L of the side of each plate is unity. The
gap between the plates is g. The dielectric constant of the external medium, ϵ,
is unity. The voltages on the upper and lower plates are $V_1 = 1$ and $V_2 = -1$,
respectively.

3.1.6.1   Capacitance evaluation from the thin plate BEM model

Results from the thin plate BEM model (with $h/L = 10^{-6}$) are compared to
those from Harrington [61] in Figure 3.7. The capacitance $C = Q/2V$, where
Q is the total charge on the top plate and V is the potential on it. $A = L^2$ is
the area of each plate. Calculation of the capacitance, of course, only requires
equation (3.5). Harrington's approach is essentially the same as the present one.

The differences are that the weakly singular integral in (3.5) is now evaluated from the method outlined in [26], and that [61] uses constant while the present work uses quadratic ($T6$) elements. The present results are believed to be more accurate than the older ones.

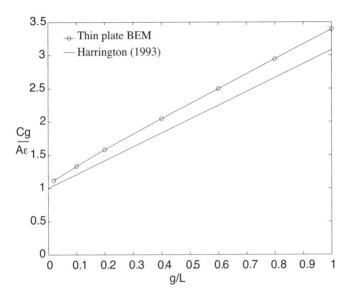

Figure 3.7: Normalized capacitance from the thin plate model compared with numerical results from [61]. Ref. [61] uses 36 constant square elements while the present work uses $16 \times 16 \times 2$ $T6$ elements on the upper surface of each plate (from [6])

3.1.6.2   Comparison of various methods

The performance of the three BEM models - the usual, the enhanced and the thin plate - are compared for the parallel plate capacitor problem (Figure 3.1) in Table 3.1. Different values of h/L are considered here while g/L is kept fixed at 0.2. Each method has different regions where it excels - the enhanced BEM being best suited for intermediate thicknesses and the thin plate BEM being best for very thin plates.

Total charge. Table 3.1 is only concerned with the total charge on the plates - therefore only with equation (3.5) of the thin plate BEM. It is first observed that there is a significant discrepancy between the usual and enhanced BEM results for $h/L = 0.1$. This is primarily due to the fact that the enhanced BEM computes nearly weakly singular integrals by the method proposed in Section 3.1.4 while the usual BEM employs standard Gaussian quadrature for evaluating these integrals. With the given mesh, the enhanced BEM delivers

	Cg/Aϵ			**Computational effort**		
h/L	Usual BEM	Enhanced BEM	Thin plate BEM	Usual BEM	Enhanced BEM	Thin plate BEM
1	2.3975			1053.9		
0.1	3.3542	2.6631	1.2351	95.7	99.7	4.2
0.05		1.7405	1.3879		100.4	4.2
0.01		1.6899	1.5611		101.2	4.2
0.005		1.6652	1.5874		100.1	4.2
0.001		1.6221	1.6094		100.1	4.2
10^{-6}			1.6200			4.2

Table 3.1: Summary of various cases for the parallel plate capacitor problem. Mesh : All cases - on each (unit) square (top and bottom) plate face $8 \times 8 \times 2$ $T6$ triangles. Usual BEM : for cube - each side face $8 \times 8 \times 2$ $T6$ triangles; for $h/L = 0.1$, each side face 8×2 $T6$ triangles. Enhanced BEM : each side face 16×2 $T3$ triangles. $g/L = 0.2$. Computational effort in cpu seconds (from [6])

$Cg/A\epsilon = 2.6631$. The same approach, with the same mesh, but with Gaussian quadrature used to evaluate nearly weakly singular integrals, yields a value of 1.9785 - a difference of over 25% ! The usual BEM model with 16 $T6$ triangles on each side face has an additional 16 midside nodes on each side face that generate additional nearly weakly singular integrals (see Figure 3.4(b)). When all these nearly weakly singular integrals are evaluated with standard Gaussian quadrature, the usual BEM yields $Cg/A\epsilon = 3.3542$. (In fact, many of the values of charge on the midside nodes on the side faces come out wrong - they are negative !) The idea here is not to deliberately downgrade the usual BEM. Rather, it is to emphasize that significant errors can arise from the evaluation of nearly weakly singular integrals by standard Gaussian quadrature. One tries, of course, to compensate for the use of Gaussian quadrature for the evaluation of nearly weakly singular integrals by using a fine mesh, or a large number of Gauss points, or both. As seen from Figure 3.5(b), however, just using a large number of Gauss points may not help very much.

The enhanced and the thin plate BEM show best agreement for $h/L = 0.001$. It is noted that even though the thin plate equations have no explicit dependence on the plate thickness h, the final results from it do depend on h because the distance from S_1^+ to S_2^+ is $g + h$ (see Figure 3.1) and the gap g is kept fixed here.

Separate charges. Perhaps even more interesting is Table 3.2, which compares how the separate charges σ^+ and σ^- (on the top plate) are calculated by the enhanced BEM and the thin plate BEM. Of course, equation (3.10) now comes into play in the thin plate BEM. As is well known, accurate determination of σ^+ and σ^- is critical for MEMS applications since the traction on a

	Enhanced BEM		Thin plate BEM	
	$h/L = 0.01$	$h/L = 0.001$	$h/L = 0.01$	$h/L = 0.001$
σ^+ at plate center	1.9972	4.5300	1.4539	1.4500
σ^- at plate center	9.8761	12.150	9.4399	9.8600
$\int_{s_1^+} \sigma^+ ds$	0.3183	0.7904	0.4303	0.4348
$\int_{s_1^-} \sigma^- ds$	1.2622	0.8182	1.1308	1.1746

Table 3.2: Comparison of enhanced BEM and thin plate BEM for the parallel plate capacitor problem. Mesh : Both cases - on each (unit) square (top and bottom) plate face $8 \times 8 \times 2$ $T6$ triangles. Enhanced BEM : each side face 16×2 $T3$ triangles. $g/L = 0.2$ (from [6])

plate, at a surface point, depends on the square of the charge density at that point. Agreement between the two methods is tolerable for $h/L = 0.01$. The enhanced BEM, however, breaks down when determining the charges separately for $h/L = 0.001$ (even though it successfully calculates the capacitance for this value of h/L). The reason for this failure is that the usual (and enhanced) BIEs, collocated at $\mathbf{x}^+$ and $\mathbf{x}^-$ in Figure 3.1, become nearly identical as $h/L \to 0$. This is analogous to the well-known failure of the usual elasticity BIE for fracture mechanics problems (see, e.g. [87, 105]), and, of course, gets worse as h/L gets smaller. The thin plate BEM performs well for $h/L = 0.001$ (and for a much lower value of h/L, see Table 3.3). This is because equation (3.5) is now only used for obtaining the sum of the charges on each plate, and (3.10) is used for obtaining the separate charges. The bottom line is that equation (3.10) is crucial for obtaining the charges separately.

mesh on unit square	σ^+ at plate center	σ^- at plate center	$Cg/A\epsilon$
$8 \times 8 \times 2$	1.455	9.913	1.615
$12 \times 12 \times 2$	1.465	9.908	1.593
$16 \times 16 \times 2$	1.470	9.907	1.583

Table 3.3: Convergence of the thin plate BEM model for the parallel plate capacitor problem for various meshes. $h/L = 10^{-6}$, $g/L = 0.2$ (from [6])

The convergence characteristics for the thin plate BEM are investigated for a super thin plate ($h/L = 10^{-6}$) in Table 3.3. It is found that the mesh dependence of the results is minimal.

Computational efficiency. Finally, Table 3.1 demonstrates huge savings in computational effort for the thin plate BEM - around a factor of 25 compared to the enhanced BEM.

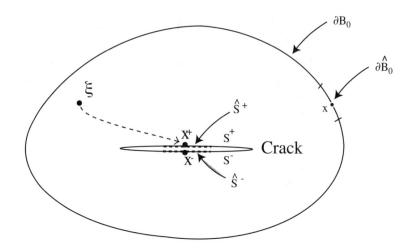

Figure 3.8: Geometry of the crack problem (from [105])

3.2   BIE for Elasticity: Cracks and Thin Shells

Regularized BIEs for linear elasticity, suitable for applications in linear elastic fracture mechanics (LEFM), are presented in this section. This is followed by a discussion of BIEs suitable for the analysis of thin elastic shells. Further details are available in [105].

3.2.1   BIES in LEFM

This section is concerned with regularization of BIEs in linear elastic fracture mechanics. The starting point is FP integrals that involve traction sums and crack opening displacements when the collocation point approaches a crack surface in the limit to the boundary (LTB) sense. The appropriate FP integrals are regularized by an addition-subtraction procedure followed by applications of Stokes' theorem to convert singular or hypersingular integrals on surfaces into regular line integrals on the bounding contours of these surfaces.

The geometry of the problem appears in Figure 3.8. The total surface of a cracked body B is $\partial B = \partial B_0 \cup S^+ \cup S^-$, where S^+ and S^- are the (coincident) upper and lower surfaces of a crack in the body. These surfaces are geometrically identical and have opposite normals at any pair of twin points $\mathbf{x}^+$ and $\mathbf{x}^-$. A singular element containing $\mathbf{x}^+$ is $\hat{S}^+$ and similarly one containing $\mathbf{x}^-$ is $\hat{S}^-$.

It is well known (e.g. Cruse [38]), that both the displacement and stress BIEs must be used appropriately in order to solve this problem by the BEM. A discussion of this, and related issues, follows.

3.2.1.1 Source point approaching the outer surface

Let an internal point $\boldsymbol{\xi}$ approach a point $\mathbf{x}$ on the outer surface ∂B_0 of the body (see Figure 3.8).

Regularized displacement BIE. The usual rigid body mode is used to regularize the integral on ∂B_0 in the displacement BIE (1.24). With $\partial B = \partial B_0 \cup S^+ \cup S^-$, a regularized form of equation (1.24) is:

$$
\begin{aligned}
0 = & \int_{\partial B_0} [U_{ik}(\mathbf{x}, \mathbf{y})\tau_i(\mathbf{y}) - T_{ik}(\mathbf{x}, \mathbf{y})(u_i(\mathbf{y}) - u_i(\mathbf{x}))]dS(\mathbf{y}) \\
& + \int_{S^+} [U_{ik}(\mathbf{x}, \mathbf{y})q_i(\mathbf{y}) - T_{ik}(\mathbf{x}, \mathbf{y})v_i(\mathbf{y})dS(\mathbf{y})
\end{aligned}
\tag{3.20}
$$

where the sum of the tractions across a crack, $\mathbf{q}$, and the crack opening displacement, $\mathbf{v}$, are defined as:

$$
q_i(\mathbf{y}) = \tau_i(\mathbf{y}^+) + \tau_i(\mathbf{y}^-)
\tag{3.21}
$$

$$
v_i(\mathbf{y}) = u_i(\mathbf{y}^+) - u_i(\mathbf{y}^-)
\tag{3.22}
$$

for twin points $\mathbf{y}^+$ and $\mathbf{y}^-$ across a crack.

Regularized stress BIE Use of both rigid body and linear displacement modes (see Lutz et al. [89] and Mukherjee et al. [99]) leads to a regularized form of the stress BIE (1.36). This is:

$$
\begin{aligned}
0 = & \int_{\partial B_0} \{D_{ijk}(\mathbf{x}, \mathbf{y})[\sigma_{kh}(\mathbf{y}) - \sigma_{kh}(\mathbf{x})]n_h(\mathbf{y}) \\
& - S_{ijk}(\mathbf{x}, \mathbf{y})[u_k(\mathbf{y}) - u_k(\mathbf{x}) - u_{k,\ell}(\mathbf{x})(y_\ell - x_\ell)]\} \, dS(\mathbf{y}) \\
& + \int_{S^+} [D_{ijk}(\mathbf{x}, \mathbf{y})q_k(\mathbf{y}) - S_{ijk}(\mathbf{x}, \mathbf{y})v_k(\mathbf{y})] \, dS(\mathbf{y})
\end{aligned}
\tag{3.23}
$$

Either of the equations (3.20) or (3.23) can be collocated at $\mathbf{x} \in \partial B_0$ when solving a LEFM problem. The simpler (3.20) is preferable and is recommended.

3.2.1.2 Source point approaching a crack surface

Let an internal point $\boldsymbol{\xi}$ approach a point $\mathbf{x}^+$ on the crack surface S^+ (see Figure 3.8).

The displacement BIE. An FP representation of the displacement BIE (1.24) with $\mathbf{x} \to \mathbf{x}^+ \in S^+$ is:

$$u_k(\mathbf{x}^+) = \int_{\partial B_0} \left[U_{ik}(\mathbf{x}^+, \mathbf{y})\tau_i(\mathbf{y}) - T_{ik}(\mathbf{x}^+, \mathbf{y})u_i(\mathbf{y}) \right] dS(\mathbf{y})$$
$$+ \oint_{S^+} \left[U_{ik}(\mathbf{x}^+, \mathbf{y})\, q_i(\mathbf{y}) - T_{ik}(\mathbf{x}^+, \mathbf{y})v_i(\mathbf{y}) \right] dS(\mathbf{y}) \quad (3.24)$$

With $\mathbf{x}^+ \in \hat{S}^+$, equation (3.24) can be written as:

$$u_k(\mathbf{x}^+) = \int_{\partial B_0} \left[U_{ik}(\mathbf{x}^+, \mathbf{y})\, \tau_i(\mathbf{y}) - T_{ik}(\mathbf{x}^+, \mathbf{y})u_i(\mathbf{y}) \right] dS(\mathbf{y})$$
$$+ \int_{S^+ - \hat{S}^+} \left[U_{ik}(\mathbf{x}^+, \mathbf{y})q_i(\mathbf{y}) - T_{ik}(\mathbf{x}^+, \mathbf{y})v_i(\mathbf{y}) \right] dS(\mathbf{y})$$
$$+ \int_{\hat{S}^+} U_{ik}(\mathbf{x}^+, \mathbf{y})q_i(\mathbf{y})\, dS(\mathbf{y})$$
$$- \int_{\hat{S}^+} T_{ik}(\mathbf{x}^+, \mathbf{y})[v_i(\mathbf{y}) - v_i(\mathbf{x})]dS(\mathbf{y})$$
$$- v_i(\mathbf{x}) \oint_{\hat{S}^+} T_{ik}(\mathbf{x}^+, \mathbf{y})dS(\mathbf{y}) \quad (3.25)$$

It is first noted that (see below (1.21)) $\Sigma_{ijk}(\mathbf{x}, \mathbf{y})n_j(\mathbf{y}) = T_{ik}(\mathbf{x}, \mathbf{y})$. Following [116, 99] (see, also, [87] and Figure 3.2):

$$\oint_{\hat{S}^+} T_{ik}(\mathbf{x}^+, \mathbf{y})dS(\mathbf{y}) = -g_{ik}(\mathbf{x}^+) - \frac{\Omega(\hat{S}^+, \mathbf{x}^+)}{4\pi}\delta_{ik} \quad (3.26)$$

where:

$$g_{ik}(\mathbf{x}^+) = \frac{1 - 2\nu}{8\pi(1 - \nu)}\epsilon_{ik\ell} \oint_{\hat{L}^+} \frac{1}{r(\mathbf{x}^+, \mathbf{y})}dz_\ell$$
$$+ \frac{\epsilon_{k\ell m}}{8\pi(1 - \nu)} \oint_{\hat{L}^+} \frac{r_{,i}(\mathbf{x}^+, \mathbf{y})r_{,\ell}(\mathbf{x}^+, \mathbf{y})}{r(\mathbf{x}^+, \mathbf{y})}dz_m \quad (3.27)$$

and the solid angle $\Omega(\hat{S}^+, \mathbf{x}^+)$, subtended by $\hat{S}^+$ at $\mathbf{x}^+$, is given by equation (3.8).

In equations (3.26 - 3.27), $\mathbf{z} = \mathbf{y} - \mathbf{x}$ and $,i \equiv \partial/\partial y_i$. Also, ϵ is the alternating tensor, δ is the Kronecker delta and the line integrals in (3.27) are evaluated in a clockwise sense when viewed from above (see Figure 3.2).

Using equations (3.26 - 3.27) and (3.8) in (3.25), a regularized form of (3.25) is obtained as:

$$\frac{1}{2}(u_k(\mathbf{x}^+) + u_k(\mathbf{x}^-)) = \int_{\partial B_0} \left[U_{ik}(\mathbf{x}^+, \mathbf{y})\tau_i(\mathbf{y}) - T_{ik}(\mathbf{x}^+, \mathbf{y})u_i(\mathbf{y}) \right] dS(\mathbf{y})$$

$$+ \int_{S^+ - \hat{S}^+} \left[U_{ik}(\mathbf{x}^+, \mathbf{y})q_i(\mathbf{y}) - T_{ik}(\mathbf{x}^+, \mathbf{y})v_i(\mathbf{y}) \right] dS(\mathbf{y})$$

$$+ \int_{\hat{S}^+} U_{ik}(\mathbf{x}^+, \mathbf{y})q_i(\mathbf{y}) dS(\mathbf{y})$$

$$- \int_{\hat{S}^+} T_{ik}(\mathbf{x}^+, \mathbf{y})[v_i(\mathbf{y}) - v_i(\mathbf{x})] dS(\mathbf{y})$$

$$+ g_{ik}(\mathbf{x}^+)v_i(\mathbf{x}) - \frac{v_k(\mathbf{x})}{4\pi} \int_0^{2\pi} \cos(\psi(\theta)) d\theta \qquad (3.28)$$

The stress BIE. An FP representation of the stress BIE with $\boldsymbol{\xi} \to \mathbf{x}^+ \in S^+$ is:

$$\sigma_{ij}(\mathbf{x}^+) = \int_{\partial B_0} \left[D_{ijk}(\mathbf{x}^+, \mathbf{y})\tau_k(\mathbf{y}) - S_{ijk}(\mathbf{x}^+, \mathbf{y})u_k(\mathbf{y}) \right] dS(\mathbf{y})$$

$$+ \fint_{S^+} \left[D_{ijk}(\mathbf{x}^+, \mathbf{y})q_k(\mathbf{y}) - S_{ijk}(\mathbf{x}^+, \mathbf{y})v_k(\mathbf{y}) \right] dS(\mathbf{y}) \qquad (3.29)$$

With $\mathbf{x}^+ \in \hat{S}^+$, equation (3.29) can be written as:

$$\sigma_{ij}(\mathbf{x}^+) = \int_{\partial B_0} \left[D_{ijk}(\mathbf{x}^+, \mathbf{y})\tau_k(\mathbf{y}) - S_{ijk}(\mathbf{x}^+, \mathbf{y})u_k(\mathbf{y}) \right] dS(\mathbf{y})$$

$$+ \int_{S^+ - \hat{S}^+} \left[D_{ijk}(\mathbf{x}^+, \mathbf{y})q_k(\mathbf{y}) - S_{ijk}(\mathbf{x}^+, \mathbf{y})v_k(\mathbf{y}) \right] dS(\mathbf{y})$$

$$+ \int_{\hat{S}^+} D_{ijk}(\mathbf{x}^+, \mathbf{y})[s_{k\ell}(\mathbf{y}) - s_{k\ell}(\mathbf{x})]n_\ell(\mathbf{y}) dS(\mathbf{y})$$

$$- \int_{\hat{S}^+} S_{ijk}(\mathbf{x}^+, \mathbf{y})[v_k(\mathbf{y}) - v_k(\mathbf{x}) - d_{kn}(\mathbf{x})(y_n - x_n^+)] dS(\mathbf{y})$$

$$- v_k(\mathbf{x}) \fint_{\hat{S}^+} S_{ijk}(\mathbf{x}^+, \mathbf{y}) dS(\mathbf{y})$$

$$+ d_{mn}(\mathbf{x}) \fint_{\hat{S}^+} \left[E_{k\ell mn} D_{ijk}(\mathbf{x}^+, \mathbf{y})n_\ell(\mathbf{y}) - S_{ijm}(\mathbf{x}^+, \mathbf{y})(y_n - x_n^+) \right] dS(\mathbf{y})$$

$$(3.30)$$

where the newly defined quantities are:

$$s_{ij}(\mathbf{x}) = \sigma_{ij}(\mathbf{x}^+) - \sigma_{ij}(\mathbf{x}^-) \qquad (3.31)$$

$$d_{ij}(\mathbf{x}) = u_{i,j}(\mathbf{x}^+) - u_{i,j}(\mathbf{x}^-) \qquad (3.32)$$

Of course, $\mathbf{s}$ is related to $\mathbf{d}$ by Hooke's law in the same manner as $\boldsymbol{\sigma}$ is related to $\nabla \boldsymbol{u}$, i.e.:

$$s_{ij} = E_{ijmn}d_{mn} \tag{3.33}$$

The last two terms on the right-hand side of equation (3.30) involve FP integrals. These are now regularized by applying Stokes' theorem.

Using the definition of S_{ijk} from (1.32), one gets:

$$\fint_{\hat{S}+} S_{ijk}(\mathbf{x}^+, \mathbf{y})dS(\mathbf{y}) = -\fint_{\hat{S}+} E_{ijmn}\Sigma_{k\ell m,n}(\mathbf{x}^+, \mathbf{y})n_\ell(\mathbf{y})dS(\mathbf{y}) \tag{3.34}$$

The hypersingular integral on the right-hand side of (3.34) can be transformed into a line integral. A careful derivation is provided in Mukherjee and Mukherjee [99], Appendix D. Applying Stokes' theorem, one has:

$$\fint_{\hat{S}+} \Sigma_{k\ell m,n}n_\ell dS = \oint_{\hat{L}+} \Sigma_{k\ell m}\epsilon_{\ell nt}dz_t \tag{3.35}$$

Finally, the desired result is:

$$\fint_{\hat{S}+} S_{ijk}(\mathbf{x}^+, \mathbf{y})dS(\mathbf{y}) = -E_{ijmn} \oint_{\hat{L}+} \Sigma_{k\ell m}(\mathbf{x}^+, \mathbf{y})\epsilon_{\ell nt}dz_t \tag{3.36}$$

where, as before, the line integral in (3.36) is evaluated in a clockwise sense when viewed from above (see Figure 3.2). This result also appears in Liu et al. [86].

The final task is to convert the last term in the right-hand side of equation (3.30) into line integrals. In order to succeed in this task, the entire term must be converted together. Otherwise, some surface integrals will survive, as in Liu et al. [86].

First define:

$$J_{ijmn} = \fint_{\hat{S}+} \left[E_{k\ell mn}D_{ijk}(\mathbf{x}^+, \mathbf{y})n_\ell(\mathbf{y}) - S_{ijm}(\mathbf{x}^+, \mathbf{y})(y_n - x_n^+) \right] dS(\mathbf{y}) \tag{3.37}$$

Using equation (3.34) and the definition of the kernel $\mathbf{D}$ in terms of $\mathbf{U}$ from equation (1.31), one can write:

$$J_{ijmn} = E_{ijpq}I_{mnpq} \tag{3.38}$$

where:

$$I_{mnpq} = -\fint_{\hat{S}+} \left[E_{k\ell mn}U_{pk,q}(\mathbf{x}^+, \mathbf{y}) - \Sigma_{m\ell p,q}(\mathbf{x}^+, \mathbf{y})(y_n - x_n^+) \right] n_\ell(\mathbf{y})dS(\mathbf{y}) \tag{3.39}$$

Integrating by parts:

$$I_{mnpq} = -\oint_{\hat{S}+} \left[E_{k\ell mn} U_{pk}(\mathbf{x}^+, \mathbf{y}) - \Sigma_{m\ell p}(\mathbf{x}^+, \mathbf{y})(y_n - x_n^+) \right]_{,q} n_\ell(\mathbf{y}) dS(\mathbf{y})$$

$$- \oint_{\hat{S}+} \Sigma_{m\ell p} \delta_{nq} n_\ell(\mathbf{y}) dS(\mathbf{y}) \equiv I^{(1)}_{mnpq} + I^{(2)}_{mnpq} \qquad (3.40)$$

Once again, the hypersingular integrals in equation (3.40) need careful treatment. The ideas presented in Mukherjee and Mukherjee [99], Appendix D, need to be employed here. (An alternative proof, using Lutz et al.'s ([89]) "tent" integrals, is available in [105]).

Define

$$F_{\ell mnp}(\mathbf{x}^+, \mathbf{y}) = E_{k\ell mn} U_{pk}(\mathbf{x}^+, \mathbf{y}) - \Sigma_{m\ell p}(\mathbf{x}^+, \mathbf{y})(y_n - x_n^+) \qquad (3.41)$$

Using Stokes' theorem in the form:

$$\int_S \left[F_{\ell mnp,q} n_\ell - F_{\ell mnp,\ell} n_q \right] ds = \oint_L F_{\ell mnp} \epsilon_{t\ell q} dz_t \qquad (3.42)$$

(where $S = \hat{S}^+ - \tilde{S}$ is a punctured surface with $\mathbf{x}^+ \in \tilde{S}$ and $L = \hat{L}^+ - \tilde{L}$ the bounding contour of S), and the fact that $F_{\ell mnp,\ell} = 0$ for $\mathbf{x}^+ \notin S$, one finally gets:

$$I^{(1)}_{mnpq}(\mathbf{x}^+) = -h_{mnpq}(\mathbf{x}^+) \qquad (3.43)$$

where:

$$h_{mnpq}(\mathbf{x}^+) = \oint_{\hat{L}+} \left[E_{k\ell mn} U_{pk}(\mathbf{x}^+, \mathbf{y}) - \Sigma_{m\ell p}(\mathbf{x}^+, \mathbf{y})(y_n - x_n^+) \right] \epsilon_{\ell qt} dz_t \quad (3.44)$$

Also, using equation (3.26) in (3.40):

$$I^{(2)}_{mnpq}(\mathbf{x}^+) = -\int_{\hat{S}+} T_{mp}(\mathbf{x}^+, \mathbf{y}) \delta_{nq} ds(\mathbf{y})$$

$$= g_{mp}(\mathbf{x}^+) \delta_{nq} + \frac{\Omega(\hat{S}^+, \mathbf{x}^+)}{4\pi} \delta_{mp} \delta_{nq} \qquad (3.45)$$

Finally, use of equations (3.33, 3.36- 3.38, 3.40, 3.43 and 3.45) in (3.30) yields a fully regularized version of (3.30). This is:

$$(1/2)[\sigma_{ij}(\mathbf{x}^+) + \sigma_{ij}(\mathbf{x}^-)] =$$

$$\int_{\partial B_0} \left[D_{ijk}(\mathbf{x}^+, \mathbf{y}) \tau_k(\mathbf{y}) - S_{ijk}(\mathbf{x}^+, \mathbf{y}) u_k(\mathbf{y}) \right] dS(\mathbf{y})$$

$$+ \int_{S^+ - \hat{S}^+} \left[D_{ijk}(\mathbf{x}^+, \mathbf{y}) q_k(\mathbf{y}) - S_{ijk}(\mathbf{x}^+, \mathbf{y}) v_k(\mathbf{y}) \right] dS(\mathbf{y})$$

$$+ \int_{\hat{S}^+} D_{ijk}(\mathbf{x}^+, \mathbf{y})[s_{k\ell}(\mathbf{y}) - s_{k\ell}(\mathbf{x})] n_\ell(\mathbf{y}) dS(\mathbf{y})$$

$$- \int_{\hat{S}^+} S_{ijk}(\mathbf{x}^+, \mathbf{y})[v_k(\mathbf{y}) - v_k(\mathbf{x}) - d_{kn}(\mathbf{x})(y_n - x_n^+)] dS(\mathbf{y})$$

$$+ E_{ijmn} v_k(\mathbf{x}) \oint_{\hat{L}^+} \Sigma_{k\ell m}(\mathbf{x}^+, \mathbf{y}) \epsilon_{\ell nt} dz_t - E_{ijpq} d_{mn}(\mathbf{x})\, h_{mnpq}(\mathbf{x}^+)$$

$$+ E_{ijpn} d_{mn}(\mathbf{x}) g_{mp}(\mathbf{x}^+) - \frac{s_{ij}(\mathbf{x})}{4\pi} \int_0^{2\pi} \cos(\psi(\theta)) d\theta \qquad (3.46)$$

3.2.2   Numerical Implementation of BIES in LEFM

Of interest here are the regularized equations (3.20), (3.23), (3.28) and (3.46). The first step in the implementation is to choose shape functions for $\mathbf{u}$ and $\boldsymbol{\tau}$ on ∂B_0 and $\mathbf{q}$ and $\mathbf{v}$ on s^+. A dot product of equation (3.46) must be taken with $\mathbf{n}(\mathbf{x}^+)$.

Next, it is necessary to obtain $u_{k,\ell}(\mathbf{x})$ $(\mathbf{x} \in \partial B_0)$ in (3.23) and $d_{k\ell}(\mathbf{x})$ $(\mathbf{x} \in s^+)$ in (3.46). The former is obtained as suggested by Lutz et al. [89]. For 3-D problems, local coordinates $\alpha_k (k = 1, 2, 3)$, are chosen at $\mathbf{x} \in \partial B_0$ such that the α_3 axis is normal and the α_1 and α_2 axes are tangential to ∂B_0 at $\mathbf{x}$. Now, tangential differentiation of the displacement shape functions provides the quantities $u_{k,\delta}, k = 1, 2, 3; \delta = 1, 2$. The remaining displacement gradients at $\mathbf{x}$ are obtained from the formulae:

$$\left. \begin{array}{l} u_{1,3} = \frac{\tau_1}{G} - u_{3,1} \\[2mm] u_{2,3} = \frac{\tau_2}{G} - u_{3,2} \\[2mm] u_{3,3} = \frac{(1-2\nu)\tau_3}{2(1-\nu)G} - \frac{\nu}{1-\nu}(u_{1,1} + u_{2,2}) \end{array} \right\} \qquad (3.47)$$

The quantities $d_{k\ell}(\mathbf{x})$ with $(\mathbf{x} \in s^+)$ require application of the above idea on both surfaces of a crack (Figure 3.9). First, in the α_k coordinate frame, one has:

$$d_{k\gamma} = v_{k,\gamma}, \quad k = 1, 2, 3 : \quad \gamma = 1, 2 \qquad (3.48)$$

The remaining components of $\mathbf{d}$ are obtained as follows:

$$\left. \begin{array}{ll} \text{On } S^+, \text{in the } \alpha_k \text{ frame :} & u_{1,3}(\mathbf{x}^+) = \tau_1^+/G - (u_3^+)_{,1} \\[2mm] \text{On } S^-, \text{in the } \beta_k \text{ frame :} & u_{1,3}(\mathbf{x}^-) = \tau_1^-/G - (u_3^-)_{,1} \\[2mm] \text{On } S^-, \text{in the } \alpha_k \text{ frame :} & -u_{1,3}(\mathbf{x}^-) = \tau_1^-/G + (u_3^-)_{,1} \end{array} \right\} \qquad (3.49)$$

Adding the first and last of equations (3.49), one has:

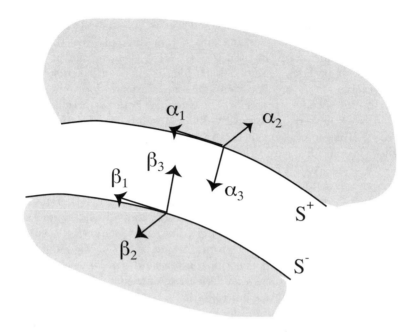

Figure 3.9: Local coordinates on crack surfaces (from [105])

$$\text{At } \mathbf{x}^+ \text{ in the } \alpha_k \text{ frame}: d_{13} = \frac{q_1}{G} - d_{31} \tag{3.50}$$

Similarly:

$$d_{23} = \frac{q_2}{G} - d_{32} \tag{3.51}$$

$$d_{33} = \frac{(1 - 2\nu)q_3}{2(1 - \nu)G} - \frac{\nu}{1 - \nu}(d_{11} + d_{22}) \tag{3.52}$$

It should be noted that in equations (3.49), $u_k^+ \equiv u_k(\mathbf{x}^+)$, $\tau_k^+ \equiv \tau_k(\mathbf{x}^+)$, and similarly at $\mathbf{x}^-$.

Finally, $\mathbf{s}$ is related to $\mathbf{d}$ by Hooke's law (3.33).

It is interesting to point out that if the surface (tangential) gradient of the displacement field (rather than its total gradient) is used to regularize the relevant HBIE, as is done, for example, in [85] - [86], then the procedure outlined in this section is no longer required. The surface gradient regularization procedure, however, leads to some extra terms in a regularized HBIE such as equation (3.23) in this chapter.

3.2.3 Some Comments on BIEs in LEFM

As has been mentioned before, equation (3.20) is the most convenient one for collocation on ∂B_0. On S^+, both the regularized displacement BIE (3.28) and the stress BIE (3.46) are, in some sense, "defective." Equation (3.28) contains $\mathbf{u}^+$ and $\mathbf{u}^-$ on the crack surfaces, but only $\mathbf{q}$ (not the individual tractions) on S^+ and on S^-. Thus, it can be used to solve for $\mathbf{q}$ given $\mathbf{u}^+$ and $\mathbf{u}^-$, but not for the important practical case of solving for the crack surface displacements given the tractions. "Overcollocation" on crack surfaces does not help since $\mathbf{q}$ alone does not specify a problem. For example, both a traction-free and a pressurized crack have $\mathbf{q} = 0$ but very different solutions. This failure of the displacement BIE for fracture mechanics problems, is, of course, well known (e.g. Cruse [38]) and has led to the search for a remedy in the form of a stress BIE, as early as at least 1975 ([17], [59]).

When one takes a dot product of the stress BIE (3.46) with $\mathbf{n}^+$, it contains the tractions $\boldsymbol{\tau}^+$, $\boldsymbol{\tau}^-$, and the crack opening displacement (COD) $\mathbf{v}$. This equation would fail if one wished to calculate the individual tractions on the crack surfaces given the COD. Fortunately, however, it is perfectly suited for the practical case of solving for $\mathbf{v}$ given $\boldsymbol{\tau}^+$ and $\boldsymbol{\tau}^-$. (As mentioned in the previous subsection, the components of the tensor $\mathbf{d}$ can be easily calculated from the tractions and the tangential derivatives of the COD (3.48, 3.50- 3.52), and $\mathbf{s}$ is related to $\mathbf{d}$ by Hooke's law (3.33)).

In summary, therefore, it is most convenient to collocate equation (3.20) on ∂B_0 and (3.46) (i.e. its dot product with $\mathbf{n}^+$) on S^+. (Lutz et al. [89] used $\mathbf{u}$ and $\boldsymbol{\tau}$ on crack surfaces as primary variables and collocated the regular BIE on one crack face and the HBIE on the other). The individual crack surface displacements $\mathbf{u}^+$ and $\mathbf{u}^-$ can then be calculated, if desired, at a postprocessing step, from (3.28) on S^+.

3.2.4 BIEs for Thin Shells

This section is concerned with BIEs and HBIEs for thin shells, especially in the limit as the thickness of a shell $\to 0$. The question of what constitutes a thin plate or shell, in the limit as the thickness $\to 0$, is an interesting one that has been discussed extensively in the literature, from both a mathematical as well as an engineering perspective.

3.2.4.1 Mathematical formulation

Regularized BIEs and HBIEs for thin shells are presented next. The geometry of the problem is shown in Figure 3.10. Again, the "tent" is chosen such that the unit outward normal to the body at $\mathbf{x}^+$ is also the outward normal to the "tent" at that point.

The situation here is quite analogous to the crack problem with the outer surface ∂B_0 in Figure 3.8 replaced by the "edge surface" S_E of the shell. It is important to retain the integral over S_E because even when the shell thickness $h \to 0$ and S_E degenerates from a surface to a curve, this integral can remain nonzero with finite applied tractions per unit length along this curve.

It has been proved in [105] that, in this case:

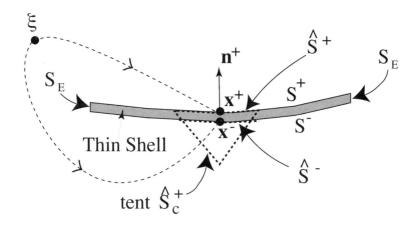

Figure 3.10: Geometry of the thin shell problem (from [105])

$$v_i(\mathbf{x}) = u_i(\mathbf{x}^+) - u_i(\mathbf{x}^-) = 0 \qquad (3.53)$$

$$s_{ij}(\mathbf{x}) = \sigma_{ij}(\mathbf{x}^+) - \sigma_{ij}(\mathbf{x}^-) = 0 \qquad (3.54)$$

Equations (3.53 - 3.54) simply state that, starting with a continuous displacement and stress field in a shell of finite thickness, the displacements and stresses on the upper and lower surfaces of the shell must approach each other as the shell thickness $h \to 0$.

Unfortunately, however, equation (3.54) implies, (in the limit $h \to 0$), a strong restriction on the applied tractions on S^+ and S^-, namely that:

$$q_k(\mathbf{x}) = \tau_k(\mathbf{x}^+) + \tau_k(\mathbf{x}^-) = 0 \qquad (3.55)$$

Strictly speaking, (3.55) appears to limit the class of thin shell problems that can be solved by this BIE approach, or, for that matter, by any other approach, since equations (3.53 - 3.55) are always true in the thin shell limit $h \to 0$. Nevertheless, it is well known, of course, that finite element shell formulations, for example, work well in practice.

There are interesting situations that satisfy (3.53 -3.54) approximately. One of these cases appears in Liu [87] and is discussed in the next subsection.

3.2.4.2 Some comments on thin shell problems

First, it is useful to examine equations (11-13) in Liu [87]. These equations are valid for $h \to 0$ in the absence of the edge integrals on S_E. Subtracting (12) from (11) immediately gives $\mathbf{\Delta u} = \mathbf{0}$ which implies that (since the matrix B^+ in [87] is nonsingular) $\Sigma \mathbf{t} = \mathbf{0}$, i.e. equations (3.53) and (3.55) of the present chapter. Therefore, equation (13) of Liu [87] (which is a restatement of his equations (11-12) in terms of displacements and tractions on the upper and lower surfaces of the shell) only admits the solutions $\mathbf{v} = \mathbf{q} = 0$ - no other solutions are possible in the thin shell limit !

One is, of course, primarily interested in solving problems for shells of small but finite thickness, i.e. application of equations (4-5) of Liu [87] (including, in general, the integral on S_E) to thin shells. (Note that (13) of [87] is the limiting form of (4-5) of [87] as $h \to 0$.) Difficulties might arise when applying equations (4-5) of [87] to shells of small but finite thickness h if the applied tractions do not satisfy (3.55), since one would then have a contradiction as $h \to 0$. Liu [87], however, has demonstrated the usefulness of his equations (4-5) for the case of a thin spherical shell under internal pressure with radius to thickness ratios of up to about 200. One reason for his success is that in his spherically symmetric numerical example the only nonzero displacement component, the radial one, is nearly uniform as a function of the radius, so that (3.53) is satisfied in the thin shell limit. Also, (3.54) is satisfied by the tangential stress components (in the thin shell limit) while the radial stress component, that violates (3.54), is small compared to the tangential ones. The shearing stress components are zero in a spherical coordinate system. Finally, the only nonzero applied traction, the radial one, violates (3.55) but its magnitude is small compared to the tangential stress components.

In view of the above discussion, the authors of this book feel that the effect of the asymptotic requirement (3.55) on the numerical performance of the standard BIE, collocated at twin points on opposite surfaces of a shell, for general thin shell problems, remains an open question that needs further investigation.

Part II

THE BOUNDARY CONTOUR METHOD

Chapter 4

LINEAR ELASTICITY

Derivations of the BCM and HBCM, for 3-D linear elasticity, together with representative numerical results for selected problems, are presented in this chapter.

4.1 Surface and Boundary Contour Equations

4.1.1 Basic Equations

A regularized form of the standard boundary integral equation (Rizzo [141]), for 3-D linear elasticity (see equation (1.26) in Chapter 1), is:

$$
\begin{aligned}
0 &= \int_{\partial B} [U_{ik}(\mathbf{x},\mathbf{y})\sigma_{ij}(\mathbf{y}) - \Sigma_{ijk}(\mathbf{x},\mathbf{y})\{u_i(\mathbf{y}) - u_i(\mathbf{x})\}] \, \mathbf{e}_j \cdot d\mathbf{S}(\mathbf{y}) \\
&\equiv \int_{\partial B} \mathbf{F}_k \cdot d\mathbf{S}(\mathbf{y}) \tag{4.1}
\end{aligned}
$$

Here, as before, ∂B is the bounding surface of a body B (B is an open set) with infinitesimal surface area $d\mathbf{S} = dS\mathbf{n}$, where $\mathbf{n}$ is the unit outward normal to ∂B at a point on it. The stress tensor is $\boldsymbol{\sigma}$, the displacement vector is $\mathbf{u}$ and $\mathbf{e}_j (j = 1, 2, 3)$ are global Cartesian unit vectors. The BEM Kelvin kernels are written in terms of (boundary) source and field points. These are given in Chapter 1 as equations (1.22) and (1.18), respectively.

The first task is to show that the integrand vector $\mathbf{F}_k$ in equation (4.1) is divergence free (except at the point of singularity $\mathbf{x} = \mathbf{y}$). Writing in component form:

$$
\mathbf{F}_k = F_{jk}\mathbf{e}_j = (\sigma_{ij}U_{ik} - \Sigma_{ijk}u_i)\mathbf{e}_j + \Sigma_{ijk}u_i(\mathbf{x})\mathbf{e}_j \tag{4.2}
$$

Taking the divergence of the above at a field point $\mathbf{y}$, one gets:

$$\nabla_{\mathbf{y}} \cdot \mathbf{F}_k = F_{jk,j} \quad = \quad (\sigma_{ij} E_{ijk} - E_{ijk}\epsilon_{ij})$$
$$+ (\sigma_{ij,j} U_{ik} - \Sigma_{ijk,j} u_i) + \Sigma_{ijk,j} u_i(\mathbf{x}) \qquad (4.3)$$

where the Kelvin strain tensor $\mathbf{E}$ and the strain field $\mathbf{e}$ are:

$$E_{ijk} = (1/2)(U_{ik,j} + U_{jk,i}), \qquad \epsilon_{ij} = (1/2)(u_{i,j} + u_{j,i}) \qquad (4.4)$$

Let $(\mathbf{u}, \boldsymbol{\sigma})$ correspond to a body force free electrostatic state with the same elastic constants as the Kelvin solution. The stress and strain tensors, σ_{ij} and ϵ_{ij}, respectively, are related to each other through the usual Hooke's law. The corresponding Kelvin stress and strain tensors Σ_{ijk} and E_{ijk}, respectively, are related by Hooke's law in exactly the same manner (see, e.g. Mukherjee [98]). As a consequence, the first expression on the right-hand side (RHS) of equation (4.3) vanishes. Also, equilibrium in the absence of body forces demands that σ_{ij} be divergence free. The corresponding Kelvin stress tensor Σ_{ijk} is also divergence free, except at the point of singularity. Therefore, the second and third expressions on the RHS of (4.3) also vanish everywhere, except at the point of singularity. Thus, $\mathbf{F}_k$ in equation (4.3) is divergence free.

The above property demonstrates the existence of vector potential functions $\mathbf{V}_k$ such that:

$$\mathbf{F}_k = \nabla \times \mathbf{V}_k \qquad (4.5)$$

As a consequence of equation (4.5), the surface integral in equation (4.1) over any open surface patch $S \in \partial B$, can be converted to a contour integral around the bounding curve C of S, by applying Stokes' theorem, i.e.:

$$\int_S \mathbf{F}_k \cdot d\mathbf{S} = \oint_C \mathbf{V}_k \cdot d\mathbf{r} \qquad (4.6)$$

Stokes' theorem is valid under very general conditions. The closed curve C and the open surface S need not be flat - they only need to be piecewise smooth.

4.1.2   Interpolation Functions

Since the vectors $\mathbf{F}_k$ contain the unknown fields $\mathbf{u}$ and $\boldsymbol{\sigma}$, interpolation functions must be chosen for these variables, and potential functions derived for each linearly independent interpolation function, in order to determine the vectors $\mathbf{V}_k$. Also, since the kernels in equation (4.1) are functions only of $z_k = y_k - x_k$ (and not of the source and field coordinates separately), these interpolation functions must also be written in the coordinates z_k in order to determine the potential vectors $\mathbf{V}_k$. Finally, these interpolation functions are global in nature and are chosen to satisfy, a priori, the Navier-Cauchy equations of equilibrium. (Such functions are called Trefftz functions). The weights, in linear combinations

$\bar{u}_{\alpha i}$	1	2	3	4	5	6	7
$i=1$	1	0	0	y_1	0	0	y_2
$i=2$	0	1	0	0	y_1	0	0
$i=3$	0	0	1	0	0	y_1	0
$\bar{u}_{\alpha i}$	8	9	10	11	12	13	14
$i=1$	0	0	y_3	0	0	y_1^2	y_2^2
$i=2$	y_2	0	0	y_3	0	$k_1 y_1 y_2$	$k_2 y_1 y_2$
$i=3$	0	y_2	0	0	y_3	0	0
$\bar{u}_{\alpha i}$	15	16	17	18	19	20	21
$i=1$	$k_2 y_1 y_2$	$k_1 y_1 y_2$	$k_1 y_1 y_3$	$k_2 y_1 y_3$	y_3^2	y_1^2	0
$i=2$	y_1^2	y_2^2	0	0	0	0	y_2^2
$i=3$	0	0	y_3^2	y_1^2	$k_2 y_1 y_3$	$k_1 y_1 y_3$	$k_1 y_2 y_3$
$\bar{u}_{\alpha i}$	22	23	24	25	26	27	
$i=1$	0	0	0	$y_2 y_3$	0	0	
$i=2$	y_3^2	$k_2 y_2 y_3$	$k_1 y_2 y_3$	0	$y_1 y_3$	0	
$i=3$	$k_2 y_2 y_3$	y_2^2	y_3^2	0	0	$y_1 y_2$	

Table 4.1: Trefftz functions for interpolating displacements. $\alpha = 1, 2, 3$ are constant, $\alpha = 4, 5, ..., 12$ are linear and $\alpha = 13, 14, ..., 27$ are quadratic. $k_1 = -4(1 - \nu)$, $k_2 = -2(1 - 2\nu)$, $k_3 = k_1 - 4$

of these interpolation functions, however, are defined piecewise on boundary elements.

Quadratic interpolation functions are used in this work. With :

$$z_k = y_k - x_k \tag{4.7}$$

one has, on a boundary element :

$$u_i = \sum_{\alpha=1}^{27} \beta_\alpha \bar{u}_{\alpha i}(y_1, y_2, y_3) = \sum_{\alpha=1}^{27} \hat{\beta}_\alpha(x_1, x_2, x_3) \bar{u}_{\alpha i}(z_1, z_2, z_3) \tag{4.8}$$

$$\sigma_{ij} = \sum_{\alpha=1}^{27} \beta_\alpha \bar{\sigma}_{\alpha ij}(y_1, y_2, y_3) = \sum_{\alpha=1}^{27} \hat{\beta}_\alpha(x_1, x_2, x_3) \bar{\sigma}_{\alpha ij}(z_1, z_2, z_3) \tag{4.9}$$

where $\bar{u}_{\alpha i}$, $\bar{\sigma}_{\alpha ij}$ (with $i = 1, 2, 3$ and $\alpha = 1, 2, .., 27$) are the interpolation functions and β_α are the weights in the linear combinations of the shape functions. Each boundary element has, associated with it, 27 constants β_α that will be related to physical variables on that element. This set β differs from one element to the next.

The displacement interpolation functions for $\alpha = 1, 2, 3$ are constants, those for $\alpha = 4, ..., 12$ are of first degree and those for $\alpha = 13, ..., 27$ are of second degree. There are a total of 27 linearly independent (vector) interpolation functions $\bar{\mathbf{u}}_\alpha$. The interpolation functions for the stresses are obtained from those for the displacements through the use of Hooke's law. The interpolation functions $\bar{u}_{\alpha i}$ and $\bar{\sigma}_{\alpha ij}$ are given in Tables 4.1 and 4.2, respectively.

$\bar{\sigma}_{\alpha ij}/G$	1	2	3	4	5	6	7
$i,j=1,1$	0	0	0	k_1/k_2	0	0	0
$i,j=1,2$	0	0	0	0	1	0	1
$i,j=1,3$	0	0	0	0	0	1	0
$i,j=2,2$	0	0	0	λ/G	0	0	0
$i,j=2,3$	0	0	0	0	0	0	0
$i,j=3,3$	0	0	0	λ/G	0	0	0
$\bar{\sigma}_{\alpha ij}/G$	8	9	10	11	12	13	14
$i,j=1,1$	λ/G	0	0	0	λ/G	$-k_1y_1$	$-4\nu y_1$
$i,j=1,2$	0	0	0	0	0	k_1y_2	$4\nu y_2$
$i,j=1,3$	0	0	1	0	0	0	0
$i,j=2,2$	k_1/k_2	0	0	0	λ/G	k_3y_1	k_1y_1
$i,j=2,3$	0	1	0	1	0	0	0
$i,j=3,3$	λ/G	0	0	0	k_1/k_2	$-4\nu y_1$	$-4\nu y_1$
$\bar{\sigma}_{\alpha ij}/G$	15	16	17	18	19	20	21
$i,j=1,1$	k_1y_2	k_3y_2	k_3y_3	k_1y_3	$-4\nu y_1$	$-k_1y_1$	$-4\nu y_2$
$i,j=1,2$	$4\nu y_1$	k_1y_1	0	0	0	0	0
$i,j=1,3$	0	0	k_1y_1	$4\nu y_1$	$4\nu y_3$	k_1y_3	0
$i,j=2,2$	$-4\nu y_2$	$-k_1y_2$	$-4\nu y_3$	$-4\nu y_3$	$-4\nu y_1$	$-4\nu y_1$	$-k_1y_2$
$i,j=2,3$	0	0	0	0	0	0	k_1y_3
$i,j=3,3$	$-4\nu y_2$	$-4\nu y_2$	$-k_1y_3$	$-4\nu y_3$	k_1y_1	k_3y_1	k_3y_2
$\bar{\sigma}_{\alpha ij}/G$	22	23	24	25	26	27	
$i,j=1,1$	$-4\nu y_2$	$-4\nu y_3$	$-4\nu y_3$	0	0	0	
$i,j=1,2$	0	0	0	y_3	y_3	0	
$i,j=1,3$	0	0	0	y_2	0	y_2	
$i,j=2,2$	$-4\nu y_2$	k_1y_3	k_3y_3	0	0	0	
$i,j=2,3$	$4\nu y_3$	$4\nu y_2$	k_1y_2	0	y_1	y_1	
$i,j=3,3$	k_1y_2	$-4\nu y_3$	$-k_1y_3$	0	0	0	

Table 4.2: Trefftz functions for interpolating stresses. $\alpha = 1, 2, 3$ are zero, $\alpha = 4, 5, ..., 12$ are constant and $\alpha = 13, 14, ..., 27$ are linear. $k_1 = -4(1 - \nu)$, $k_2 = -2(1 - 2\nu)$, $k_3 = k_1 - 4$

It is easy to show that the coordinate transformation (4.7) results in the constants $\hat{\beta}_j$ being related to the constants β_α as follows :

$$\hat{\beta}_i = \sum_{\alpha=1}^{27} S_{i\alpha}(x_1, x_2, x_3)\beta_\alpha , \quad i = 1, 2, 3 \tag{4.10}$$

$$\hat{\beta}_k = \sum_{\alpha=1}^{27} R_{n\alpha}(x_1, x_2, x_3)\beta_\alpha \ , \quad k = 4, 5, ...12, \quad n = k - 3 \qquad (4.11)$$

$$\hat{\beta}_\alpha = \beta_\alpha \ , \qquad\qquad \alpha = 13, 14, .., 27 \qquad (4.12)$$

where

$$S_{i\alpha} \;=\; \bar{u}_{\alpha i}(x_1, x_2, x_3) \ , \qquad i = 1, 2, 3, \qquad \alpha = 1, 2, ..., 27$$

$$R_{k\alpha} \;=\; \left. \frac{\partial \bar{u}_{\alpha\ell}(y_1, y_2, y_3)}{\partial y_j} \right|_{(x_1, x_2, x_3)} , \quad k = 1, 2, .., 9, \quad \alpha = 1, 2, ..., 27$$

with $j = 1 + \lfloor \frac{k-1}{3} \rfloor$ and $\ell = k - 3j + 3$. Here, the symbol $\lfloor n \rfloor$, called the floor of n, denotes the largest integer less than or equal to n.

It is useful to note that the matrices $\mathbf{S}$ and $\mathbf{R}$ are functions of only the source point coordinates (x_1, x_2, x_3).

4.1.3   Boundary Elements

The BCM is a perfectly general approach that can be used to solve any well-posed problem in linear elasticity. A departure from the usual BEM, however, is that a set of primary physical variables a_k are first chosen at appropriate points on a boundary element. Some of these would be specified as boundary conditions and the rest would be unknown in a given problem. The first step in the BCM solution procedure is to determine the unspecified primary physical variables in terms of those that are prescribed as boundary conditions. Once all the primary physical variables are known, the rest of the physical variables (the secondary ones) are obtained at a simple postprocessing step. Also, unlike in the standard BEM, it is particularly easy to obtain surface variables, such as stresses and curvatures, in the BCM. Surface stresses are discussed in Section 4.1.7.

The number of primary physical variables on a boundary element must match the number of artificial variables β_k associated with it, in order that the transformation matrix T relating the vectors $\mathbf{a}$ and $\boldsymbol{\beta}$ on element m is square. This relationship is expressed as:

$$\overset{m}{a} \;=\; \overset{mm}{T\beta} \qquad (4.13)$$

Of course, the matrix T must be invertible. The issue of invertibility is discussed in Nagarajan et al. [116].

The CIM9 boundary element, shown in Figure 4.1(a), is used in the present work. The displacement $\mathbf{u}$ is the primary physical variable at the three corner nodes C_i and the three midside nodes M_i, while tractions are primary variables at the internal nodes I_i. Thus, there are a total of 27 primary physical variables.

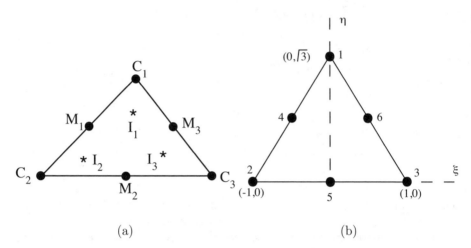

Figure 4.1: (a) The CIM9 boundary element. (b) Intrinsic coordinates (from [109])

The BCM equations are collocated at the six peripheral nodes as well as at the centroid of the element. In a typical discretization procedure, some of the peripheral nodes may lie on corners or edges, while the internal nodes are always located at regular points where the boundary ∂B is locally smooth. It is of obvious advantage to have to deal only with displacement components, that are always continuous, on edges and corners, while having traction components only at regular boundary points. This approach eliminates the well-known problems associated with modeling of corners and edges in the usual BEM.

The boundary elements, in general, are curved (quadratic) with their shapes defined by the six points $C_k, M_k,\ k = 1, 2, 3.$ (see Figure 4.1). The relative coordinates z_i (see equation (4.7)) of a point on one of the sides of a triangle are written as:

$$z_i = N_k(\xi, \eta) z_i^k, \quad i = 1, 2, 3; \ k = 1, 2, ..., 6 \tag{4.14}$$

where z_i^k are the relative coordinates of the peripheral nodes $1, 2, .., 6$ on the CIM9 element (see Figure 4.1), and ξ and η are intrinsic coordinates. The shape functions are:

$$
\begin{aligned}
N_k &= (2L_k - 1)L_k, \ k = 1, 2, 3; \text{ no sum over } k \\
N_4 &= 4L_1L_2, \ N_5 = 4L_2L_3, \ N_6 = 4L_1L_3
\end{aligned} \tag{4.15}
$$

with:

$$L_1 = \eta/\sqrt{3}$$
$$L_2 = (1/2)(1 - \xi) - \eta/(2\sqrt{3})$$
$$L_3 = (1/2)(1 + \xi) - \eta/(2\sqrt{3}) \tag{4.16}$$

The reference triangle in intrinsic coordinates is shown in Figure 4.1(b).

The unit outward normal to a boundary element, at a point on it, is given by:

$$\mathbf{n} = \left(\frac{\partial \mathbf{r}}{\partial \xi} \times \frac{\partial \mathbf{r}}{\partial \eta} \right) \Big/ \left| \frac{\partial \mathbf{r}}{\partial \xi} \times \frac{\partial \mathbf{r}}{\partial \eta} \right| \tag{4.17}$$

where $\mathbf{r} = z_i \mathbf{e}_i$. It is important to point out that the elements of the transformation matrix T in equation (4.13) contain the components of the normal $\mathbf{n}$ at the points I_1, I_2, I_3 of the CIM9 element shown in Figure 4.1(a).

4.1.4   Vector Potentials

The homogeneous nature of the Kelvin kernels is exploited in deriving the potential functions. From equations (1.22) and (1.18), it is clear that both Σ_{ijk} and U_{ik} are ratios of homogeneous polynomials and are, therefore, homogeneous. Here, Σ is of degree -2 and $\mathbf{u}$ is of degree -1. If an interpolation function is of degree n, then the resulting force vector $\mathbf{F}_{\alpha k}$ (which is $\mathbf{F}_k$ corresponding to the α-th interpolation function) is of degree $n - 2$. In this work, interpolation functions with $n = 0, 1, 2$ are used (see Tables 4.1 and 4.2).

4.1.4.1   The nonsingular case : $n \neq 0$

The homogeneous nature of the Kelvin kernels greatly facilitates the use of an inversion integral to calculate the inverse curl of a given vector field of zero divergence (see, e.g. [69]). Thus:

$$\mathbf{F}_{\alpha k} = \nabla \times \mathbf{V}_{\alpha k} \quad \Rightarrow \quad \mathbf{V}_{\alpha k}(z_1, z_2, z_3) = \left[\int_0^1 t \mathbf{F}_{\alpha k}(t z_1, t z_2, t z_3) dt \right] \times \mathbf{r} \tag{4.18}$$

where $\mathbf{r} = z_i \mathbf{e}_i$.

Because of the homogeneous nature of $\mathbf{F}_{\alpha k}$:

$$\mathbf{F}_{\alpha k}(t z_1, t z_2, t z_3) = t^{n-2} \mathbf{F}_{\alpha k}(z_1, z_2, z_3) \tag{4.19}$$

Therefore, for the nonsingular case $n \neq 0$ equations (4.18, 4.19) yield:

$$\mathbf{V}_{\alpha k}(z_1, z_2, z_3) = (1/n) \mathbf{F}_{\alpha k}(z_1, z_2, z_3) \times \mathbf{r} \tag{4.20}$$

4.1.4.2   The singular case : n=0

The singular case ($n = 0$) corresponds to constant displacement interpolation functions $\bar{u}_{\alpha i} = \delta_{\alpha i}$, $\alpha = 1, 2, 3$; $i = 1, 2, 3$, (where δ is the Kronecker delta - see Table 4.1). Referring to equation (4.1), one must now deal with the term $\Sigma_{\alpha jk}$ given in equation (1.22). The expression for $\Sigma_{ijk}\mathbf{e}_j$ can be partitioned into three terms, each of which are divergence free [116]. One writes:

$$-\Sigma_{\alpha jk}\mathbf{e}_j = \frac{1}{8\pi(1-\nu)r^2}\left[3r_{,\alpha}r_{,j}r_{,k} - r_{,j}\delta_{\alpha k}\right]\mathbf{e}_j$$
$$+\frac{1-2\nu}{8\pi(1-\nu)r^2}\left[r_{,\alpha}\delta_{jk} - r_{,k}\delta_{\alpha j}\right]\mathbf{e}_j + \frac{r_{,j}\delta_{\alpha k}}{4\pi r^2}\mathbf{e}_j \quad (4.21)$$

Each of the above three terms on the right-hand side of equation (4.21) are divergence free and can be written as the curl of a potential function. Corresponding to the first two terms, one has [116]:

$$\frac{1}{r^2}\left[3r_{,\alpha}r_{,j}r_{,k} - r_{,j}\delta_{\alpha k}\right]\mathbf{e}_j = \nabla \times \left[\epsilon_{kmj}\frac{r_{,\alpha}r_{,m}}{r}\mathbf{e}_j\right] \quad (4.22)$$

$$\frac{1}{r^2}\left[r_{,\alpha}\delta_{jk} - r_{,k}\delta_{\alpha j}\right]\mathbf{e}_j = \nabla \times \left[\frac{\epsilon_{\alpha km}}{r}\mathbf{e}_m\right] \quad (4.23)$$

where ϵ is the alternating tensor.

It is well known that the solid angle Ω subtended at the source point $\mathbf{x}$ by the open surface S has the expression:

$$\Omega = \int_S \frac{r_{,j}}{r^2}\mathbf{e}_j \cdot d\mathbf{S} = \int_S \frac{\mathbf{r}.d\mathbf{S}}{r^3} \quad (4.24)$$

Therefore, the surface integral of the third term on the right-hand side of (4.21) over S equals $(\Omega/4\pi)\delta_{\alpha k}$. While it is possible to convert the surface integral in (4.24) to a line integral [116], use of this line integral has proved to lack robustness in general numerical computations involving 3-D bodies of complex shape. Therefore, the solid angle Ω is computed as a surface integral according to equation (4.24). *This is the only surface integral that is ever computed in the BCM*. Fortunately, the solid angle is a purely geometrical quantity that can be computed easily as a surface integral in a robust fashion. It is also noted here that algebraic expressions exist for the solid angle for the special case when S is a plane.

4.1.5   Final BCM Equations

Use of the interpolating functions for displacement and stress from Tables 4.1 and 4.2 respectively, together with equations (4.7 - 4.13) and (4.20 - 4.24) transforms the regularized BIE (4.1) into a regularized boundary contour equation (BCE) that can be collocated (as in the usual BEM) at any boundary (surface) point - including those on edges and corners, as long as the displacement is continuous there. This equation is:

$$
\begin{aligned}
0 \;=\; & \frac{1}{2}\sum_{m=1}^{M}\sum_{\alpha=13}^{27}\left[\oint_{L_m}\left(\overline{\sigma}_{\alpha ij}U_{ik}-\overline{u}_{\alpha i}\Sigma_{ijk}\right)\epsilon_{jnt}z_n dz_t\right]\left[\overset{m}{T}^{-1}\overset{m}{a}\right]_{\alpha} \\
& +\sum_{m=1}^{M}\sum_{\alpha=4}^{12}\left[\oint_{L_m}\left(\overline{\sigma}_{\alpha ij}U_{ik}-\overline{u}_{\alpha i}\Sigma_{ijk}\right)\epsilon_{jnt}z_n dz_t\right]\left[R\,\overset{m}{T}^{-1}\overset{m}{a}\right]_{\alpha-3} \\
& +\sum_{\substack{m=1\\m\notin S}}^{M}\sum_{\alpha=1}^{3}\left[\oint_{L_m}D_{\alpha jk}dz_j\right]\left[S\left(\overset{m}{T}^{-1}\overset{m}{a}-\overset{P}{T}^{-1}\overset{P}{a}\right)\right]_{\alpha}
\end{aligned}
\tag{4.25}
$$

with

$$
\begin{aligned}
\oint_{L_m}D_{\alpha jk}dz_j \;=\; & -\int_{S_m}\Sigma_{\alpha jk}\mathbf{e}_j\cdot d\mathbf{S} \\
=\; & \frac{1}{8\pi(1-\nu)}\oint_{L_m}\epsilon_{kij}\frac{r_{,\alpha}r_{,i}}{r}dz_j \\
& +\frac{1-2\nu}{8\pi(1-\nu)}\oint_{L_m}\epsilon_{\alpha kj}\frac{1}{r}dz_j+\frac{\Omega}{4\pi}\delta_{\alpha k}
\end{aligned}
\tag{4.26}
$$

Here L_m is the bounding contour of the surface element S_m. Also, $\overset{m}{T}$ and $\overset{m}{a}$ are the transformation matrix and primary physical variable vectors on element m, $\overset{P}{T}$ and $\overset{P}{a}$ are the same quantities evaluated on any element that belongs to the set S of elements that contain the source point $\mathbf{x}$, and ϵ_{ijk} is the usual alternating symbol.

This method of integrating Σ_{ijk} has been presented before, using spherical coordinates, by Ghosh and Mukherjee [48]. Also, Mantič [94] has independently derived the result in equation (4.26) with the use of the tangential differential operator:

$$
D_{ij}(f(\mathbf{y}))\equiv n_i(\mathbf{y})f_{,j}(\mathbf{y})-n_j(\mathbf{y})f_{,i}(\mathbf{y})
\tag{4.27}
$$

The equations are assembled by making use of the fact that displacements are continuous across elements. The final result is:

$$
Ka \;=\; 0
\tag{4.28}
$$

which is written as:

$$
Ax \;=\; By
\tag{4.29}
$$

where $\mathbf{x}$ contains the unknown and $\mathbf{y}$ the known (from the boundary conditions) values of the primary physical variables on the surface of the body. Once these equations are solved, the vector **a** is completely known. Now, at a postprocessing step, $\overset{m}{\beta_\alpha}$ can be easily obtained on each boundary element from equation (4.13).

4.1.6   Global Equations and Unknowns

The global system equation (4.29) generally leads to a rectangular matrix A. The system of linear equations is usually overdetermined but always consistent. A count of equations and unknowns is given below.

For any general polyhedron, Euler's theorem states that:

$$F + V = E + 2 \qquad (4.30)$$

where F is the number of faces (here the number of elements), V is the number of vertices (here the number of corner nodes) and E is the number of edges (here $1.5F$ for triangular elements). Thus, one has:

$$\text{Number of corner nodes} = \frac{1}{2} \text{ Number of elements} + 2 \qquad (4.31)$$

Also, a CIM9 element has 1.5 midside nodes.

One is now in a position to count the number of (vector) equations and unknowns in the global system (4.29). In a CIM9 element, the BCM equations are enforced at all the peripheral nodes and also at the centroid of the element. Thus, for M boundary elements, one has M equations at the centroids, $1.5M$ equations at the midside nodes and $0.5M + 2$ at the corner nodes, for a total of $3M + 2$ equations. For a Dirichlet problem, in which all the displacements are prescribed on ∂B, there are a total of $3M$ vector unknowns - the tractions at the nodes interior to the elements. One therefore has two extra vector (six extra scalar) equations. This is the worst-case scenario in the sense that for mixed boundary value problems the number of equations remain the same while the number of unknowns decreases. For example, for a Neumann problem in which all the tractions are prescribed on ∂B, one only has $2M + 2$ (vector) unknowns (displacements at the peripheral nodes of the elements). Of course, a Neumann problem is ill-posed since rigid body displacements of the body are not constrained.

In summary, use of the CIM9 element results in overdetermined, consistent linear systems for well-posed problems in linear elasticity.

4.1.7   Surface Displacements, Stresses, and Curvatures

A very useful consequence of using global shape functions is that, once the standard BCM is solved, it is very easy to obtain the displacements, stresses and curvatures at a regular off-contour boundary point (ROCBP) on the bounding surface of a body. Here, a point at an edge or corner is called an irregular point while at a regular point the boundary is locally smooth. Also, a regular boundary point can lie on or away from a boundary contour. The former is called a regular contour point (RCP), the latter a regular off-contour boundary point. A point inside a body is called an internal point.

First, one obtains $\overset{m}{\beta_\alpha}$ from equation (4.13), then uses equations (4.10 and 4.11) to get $\hat{\beta}_\alpha^m$, $\alpha = 1, 2, .., 12$. The curvatures, which are piecewise constant on each boundary element, are obtained by direct differentiation of equation (4.8). Finally, one has the following results.

4.1.7.1   Surface displacements

$$[u_i(\mathbf{x})] = \begin{bmatrix} \hat{\beta}_1 \\ \hat{\beta}_2 \\ \hat{\beta}_3 \end{bmatrix}_{\mathbf{x}} \tag{4.32}$$

4.1.7.2   Surface stresses

$$[u_{i,j}(\mathbf{x})] = \begin{bmatrix} \hat{\beta}_4 & \hat{\beta}_7 & \hat{\beta}_{10} \\ \hat{\beta}_5 & \hat{\beta}_8 & \hat{\beta}_{11} \\ \hat{\beta}_6 & \hat{\beta}_9 & \hat{\beta}_{12} \end{bmatrix}_{\mathbf{x}} \tag{4.33}$$

4.1.7.3   Surface curvatures

$$\left[\frac{\partial^2 u_1}{\partial x_i \partial x_j}\right] = \begin{bmatrix} 2(\beta_{13} + \beta_{20}) & k_2\beta_{15} + k_1\beta_{16} & k_1\beta_{17} + k_2\beta_{18} \\ & 2\beta_{14} & \beta_{25} \\ \text{symmetric} & & 2\beta_{19} \end{bmatrix} \tag{4.34}$$

$$\left[\frac{\partial^2 u_2}{\partial x_i \partial x_j}\right] = \begin{bmatrix} 2\beta_{15} & k_1\beta_{13} + k_2\beta_{14} & \beta_{26} \\ & 2(\beta_{16} + \beta_{21}) & k_2\beta_{23} + k_1\beta_{24} \\ \text{symmetric} & & 2\beta_{22} \end{bmatrix} \tag{4.35}$$

$$\left[\frac{\partial^2 u_3}{\partial x_i \partial x_j}\right] = \begin{bmatrix} 2\beta_{18} & \beta_{27} & k_2\beta_{19} + k_1\beta_{20} \\ & 2\beta_{23} & k_1\beta_{21} + k_2\beta_{22} \\ \text{symmetric} & & 2(\beta_{17} + \beta_{24}) \end{bmatrix} \tag{4.36}$$

with $k_1 = -4(1 - \nu)$ and $k_2 = -2(1 - 2\nu)$

Equation (4.33) can be used to find the displacement gradients at $\mathbf{x}$. Hooke's law would then give the stress $\sigma_{ij}(\mathbf{x})$. An alternative approach is to use equation (4.9) together with all the β_α on an element.

It should be noted that the simple approach, described above, cannot be used to find internal stresses since the constants $\overset{m}{\beta_\alpha}$ are only meaningful on the boundary of a body. Therefore, an internal point representation of the differentiated BCE, for the internal displacement gradients $u_{i,j}(p)$, is necessary (see Section 4.3). It is also of interest to examine the limiting process of a differentiated BCE as an internal point ξ (also denoted as p) approaches a boundary point $\mathbf{x}$ (also denoted as P). This issue is of great current interest in the BEM community in the context of the standard and hypersingular BIE (HBIE - see, for example, [92], [39], [93], [111]). Further, the hypersingular BCE (HBCE) must be understood if one wishes to collocate the HBCE as the primary integral equation, as may be necessary, for example, in applications such as fracture mechanics, symmetric Galerkin formulations or adaptive analysis. This topic is the subject of Section 4.2 of this chapter.

4.2   Hypersingular Boundary Integral Equations

4.2.1   Regularized Hypersingular BIE

A hypersingular boundary integral equation (HBIE) can be obtained by differentiating the standard BIE *at an internal point*, with respect to the coordinates of this internal source point. A regularized version of this equation, containing, at most, weakly singular integrals (see equation (1.38) in Chapter 1) is:

$$
\begin{aligned}
0 = & \int_{\partial B} U_{ik,n}(\mathbf{x},\mathbf{y}) \left[\sigma_{ij}(\mathbf{y}) - \sigma_{ij}(\mathbf{x})\right] n_j(\mathbf{y}) dS(\mathbf{y}) \\
& - \int_{\partial B} \Sigma_{ijk,n}(\mathbf{x},\mathbf{y}) \left[u_i(\mathbf{y}) - u_i(\mathbf{x}) - u_{i,\ell}(\mathbf{x})\left(y_\ell - x_\ell\right)\right] n_j(\mathbf{y}) dS(\mathbf{y})
\end{aligned}
$$

$$\mathbf{x}, \mathbf{y} \in \partial B \qquad\qquad (4.37)$$

Martin et al. [93] - Appendix II2, p. 905, (see, also, [111]) have proved that (4.37) can be collocated even at an edge or corner point $\mathbf{x}$ on the surface of a 3-D body, provided that the displacement and stress fields in (4.37) satisfy certain smoothness requirements. These smoothness requirements are discussed in Section 1.4.4 in Chapter 1. Please note that in (4.37) above, $, n \equiv \partial y_n$, not the normal derivative.

4.2.2   Regularized Hypersingular BCE

The regularized HBIE (4.37) can be converted to a regularized hypersingular boundary contour equation (HBCE). Details are available in [99] and are given below.

The first task at hand is to transform equation (4.37) into a boundary contour form. The integrands in equation (4.37), without $n_j(\mathbf{y})dS(\mathbf{y})$, are first evaluated at an internal field point q very near Q (i.e. on a surface $\partial \hat{B}$ inside the body, very near and parallel to ∂B), the derivatives are transferred from the kernels to the quantities inside the square brackets by the product rule, and then the limit $q \to Q$ is taken again. This is possible since the integrals in equation (4.37) are regular. The result is:

$$0 = \int_{\partial B} \left[U_{ik}(\mathbf{x},\mathbf{y}) \left[\sigma_{ij}(\mathbf{y}) - \sigma_{ij}(\mathbf{x}) \right] - \Sigma_{ijk}(\mathbf{x},\mathbf{y}) \left(u_i(\mathbf{y}) - u_i^{(L)} \right) \right]_{,n} n_j(\mathbf{y}) dS(\mathbf{y})$$

$$- \int_{\partial B} \left[U_{ik}(\mathbf{x},\mathbf{y}) \sigma_{ij,n}(\mathbf{y}) - \Sigma_{ijk}(\mathbf{x},\mathbf{y}) \left[u_{i,n}(\mathbf{y}) - u_{i,n}(\mathbf{x}) \right] \right] n_j(\mathbf{y}) dS(\mathbf{y})$$

$$\mathbf{x}, \mathbf{y} \in \partial B \qquad (4.38)$$

where

$$u_i^{(L)} = u_i(\mathbf{x}) + u_{i,\ell}(\mathbf{x})(y_\ell - x_\ell)$$

Note also that:

$$u_{i,n}^{(L)} = u_{i,n}(\mathbf{x})$$

Therefore, the linear displacement field $u_i^{(L)}$ gives the stress field $\sigma_{ij}(\mathbf{x})$, so that the stress field $\sigma_{ij}(\mathbf{y}) - \sigma_{ij}(\mathbf{x})$ is obtained from the displacement field $u_i(\mathbf{y}) - u_i^{(L)}$.

Using the identity [116]

$$\mathbf{v}_{,n} = \nabla \times (\mathbf{v} \times \mathbf{e}_n) \qquad (4.39)$$

(which is valid if the vector field $\mathbf{v}$ is divergence-free) and Stokes' theorem, the first integral in equation (4.38), over S_m, is converted to the contour integral:

$$I_1 = \oint_{L_m} \left[U_{ik}(\mathbf{x},\mathbf{y}) \left[\sigma_{ij}(\mathbf{y}) - \sigma_{ij}(\mathbf{x}) \right] - \Sigma_{ijk}(\mathbf{x},\mathbf{y}) \left[u_i(\mathbf{y}) - u_i^{(L)} \right] \right] \epsilon_{jnt} dz_t$$

$$(4.40)$$

An explicit form of equation (4.40) is derived in Appendix A of this chapter. Next, it is noted from equations (4.32 and 4.10) that:

$$u_{i,n}(\mathbf{x}) = \left[S_{,N} \begin{matrix} P \\ \beta \end{matrix} \right]_i \qquad (4.41)$$

where , $N \equiv \partial/\partial x_n$.

Further, as proved in Appendix B of this chapter:

$$u_{i,n}(\mathbf{y}) = \left[S_{,N} \begin{matrix} m \\ \beta \end{matrix} \right]_i + \sum_{\alpha=4}^{12} \left[R_{,N} \begin{matrix} m \\ \beta \end{matrix} \right]_{\alpha-3} \overline{u}_{\alpha i}(z_1, z_2, z_3) \qquad (4.42)$$

Now, the second integral (called I_2) in equation (4.38) is written as:

$$I_2 = - \int_{\partial B} \left[U_{ik}(\mathbf{x},\mathbf{y}) \sigma_{ij,n}(\mathbf{y}) - \Sigma_{ijk}(\mathbf{x},\mathbf{y}) u_{i,n}(\mathbf{y}) \right] n_j(\mathbf{y}) dS(\mathbf{y})$$

$$- u_{i,n}(\mathbf{x}) \int_{\partial B} \Sigma_{ijk}(\mathbf{x},\mathbf{y}) n_j(\mathbf{y}) dS(\mathbf{y}) \qquad (4.43)$$

The next steps involve writing $\partial B \equiv \cup S_m$, separating the constant and linear parts of $u_{i,n}$ in the first integrand above, and using equation (4.42). This sets the stage for converting I_2 into two contour integrals. Details are given in Appendix C of this chapter.

The final result is a contour integral form (HBCE) of the regularized HBIE (4.37). This equation is valid at any point on the boundary ∂B as long as the stress is continuous there. This includes edge, corner and regular points, that lie on or off contours.

The regularized hypersingular boundary contour equation (HBCE) is:

$$
\begin{aligned}
0 \;=\; & -\sum_{m=1}^{M}\sum_{\alpha=13}^{27}\left[\oint_{L_m}(\overline{\sigma}_{\alpha ij}U_{ik}-\overline{u}_{\alpha i}\Sigma_{ijk})\,\epsilon_{jnt}dz_t\right]\left[\overset{m}{T^{-1}\overset{m}{a}}\right]_\alpha \\[2mm]
& +\sum_{m=1}^{M}\sum_{\alpha=4}^{12}\left[\oint_{L_m}(\overline{\sigma}_{\alpha ij}U_{ik}-\overline{u}_{\alpha i}\Sigma_{ijk})\,\epsilon_{jst}z_s dz_t\right]\left[R_{,N}\,\overset{m}{T^{-1}\overset{m}{a}}\right]_{\alpha-3} \\[2mm]
& -\sum_{\substack{m=1\\ m\notin S}}^{M}\sum_{\alpha=4}^{12}\left[\oint_{L_m}(\overline{\sigma}_{\alpha ij}U_{ik}-\overline{u}_{\alpha i}\Sigma_{ijk})\,\epsilon_{jnt}dz_t\right]\left[R\left(\overset{m}{T^{-1}\overset{m}{a}}-\overset{P}{T^{-1}\overset{P}{a}}\right)\right]_{\alpha-3} \\[2mm]
& +\sum_{\substack{m=1\\ m\notin S}}^{M}\sum_{\alpha=1}^{3}\left[\oint_{L_m}D_{\alpha jk}dz_j\right]\left[S_{,N}\left(\overset{m}{T^{-1}\overset{m}{a}}-\overset{P}{T^{-1}\overset{P}{a}}\right)\right]_\alpha \\[2mm]
& +\sum_{\substack{m=1\\ m\notin S}}^{M}\sum_{\alpha=1}^{3}\left[\oint_{L_m}\Sigma_{\alpha jk}\epsilon_{jnt}dz_t\right]\left[S\left(\overset{m}{T^{-1}\overset{m}{a}}-\overset{P}{T^{-1}\overset{P}{a}}\right)\right]_\alpha \qquad (4.44)
\end{aligned}
$$

where, as before, S is the set of boundary elements that contains the source point $\mathbf{x}$. The derivatives $R_{,N}$ and $S_{,N}$ in (4.44) are taken with respect to the source point coordinates x_n. In equation (4.44), the integrands in the first two terms are regular ($O(1)$). The third and fourth (potentially strongly singular, $O(1/r)$) as well as the fifth (potentially hypersingular, $O(1/r^2)$) need to be evaluated only on nonsingular elements.

4.2.3   Collocation of the HBCE at an Irregular Surface Point

Martin et al. [93] have stated the requirements for collocating a regularized HBIE at an irregular point on ∂B. This matter is discussed in Section 1.4.4 in Chapter 1 of this book. It has been proved in Mukherjee and Mukherjee [111] that the BCE interpolation functions given in equations (4.8) and (4.9) satisfy, a priori, all these smoothness requirements for collocation of the HBCE (4.44) at an irregular surface point. This proof is repeated below.

It is first noted that the interpolation functions in (4.8 - 4.9) have both a global (they are initially defined as functions of $\mathbf{y}$) as well as a local (the weights

β_k are only defined piecewise on the boundary elements) character. Consider a singular boundary element containing the source and field points P and Q, with P an irregular point on ∂B. Let this element be any one of the smooth pieces of ∂B that meet at P (see Section 1.4.4 in Chapter 1). From equations (4.8 - 4.9), it is easy to show that :

$$
\begin{aligned}
u_i(Q) - u_i^L(Q; P) &= u_i(Q) - u_i(P) - u_{i,j}(P)[y_j(Q) - x_j(P)] \\
&= \sum_{\alpha=13}^{27} \beta_\alpha \bar{u}_{\alpha i}(\mathbf{z}) = O(r^2) \qquad (4.45)
\end{aligned}
$$

$$
\sigma_{ij}(Q) - \sigma_{ij}(P) = \sum_{\alpha=13}^{27} \beta_\alpha \bar{\sigma}_{\alpha ij}(\mathbf{z}) = O(r) \qquad (4.46)
$$

where $r = |\mathbf{y}(Q) - \mathbf{x}(P)| = |\mathbf{z}|$. The last equalities in the above equations are true in view of the fact that the shape functions $\bar{u}_{\alpha i}$ and $\bar{\sigma}_{\alpha ij}$ are quadratic and linear, respectively, in z_k, for $\alpha = 13, 14, ..., 27$ (see Tables 4.1 and 4.2). Note that these weights β_α belong to the element containing P and Q and are unique on that element (see below equation (4.9)).

As an aside, it is interesting to connect with Toh and Mukherjee ([168]-p.2304) where, for the same problem, the requirement $|\nabla \mathbf{u}(Q) - \nabla \mathbf{u}(P)| = O(r^\alpha)$ is prescribed as $r \to 0$. It is easy to show that (Mukherjee and Mukherjee [99]), for the BCM shape functions on a singular element:

$$
u_{i,j}(Q) - u_{i,j}(P) = \sum_{\alpha=4}^{12} [R_{,J}(\mathbf{x}) \; \beta]_{\alpha-3} \bar{u}_{\alpha i}(\mathbf{z}) = O(r) \qquad (4.47)
$$

since $\bar{u}_{\alpha i}(\mathbf{z})$ is linear in $\mathbf{z}$ for $\alpha = 4, 5, ..., 12$.

In view of equations (4.45 - 4.46), conditions (iii-iv) in Box 1.1 (in Section 1.4.4 in Chapter 1) are satisfied a priori by the BCM shape functions defined by equations (4.8 - 4.9). Satisfaction of condition (ii(b)) on ∂B follows from equation (4.46). The conditions inside B ((i) and (ii(a))), of course, have nothing to do with BEM shape functions that are only defined on the bounding surface ∂B, but rather with the boundary element method itself. The BCM is derived from the BEM and it satisfies these internal point conditions in the same way as does the BEM. (As a bonus, the BCM shape functions satisfy the Navier-Cauchy equilibrium equations of linear elasticity a priori (see Section 4.1.2)), although weights are not defined at points $p \in B$). Please note that the above arguments have been made for the "worst-case scenario" of the collocation point $P \in \partial B$ being an *irregular* point. Of course, these arguments also go through for regular points (on or off contour) on ∂B.

In view of the above, all the conditions (i-iv) in Box 1.1 are satisfied a priori by the BCM, and there is no need to consider "relaxed smoothness requirements" in this method. It is worth repeating again that it is extremely difficult to find, in general, BEM shape functions (for 3-D elasticity problems) that satisfy conditions (ii(b)-iv) a priori. The primary reason for this is that BEM shape functions are defined only on the bounding surface of a body, while the BCM ones are defined in B (although the weights are defined only on ∂B).

4.3 Internal Displacements and Stresses

At this stage, it is a simple matter to derive equations for displacements and stresses at a point inside a body. The general equations are derived in [103] and equations for internal points close to the bounding surface are derived in [104]. They are given below.

4.3.1 Internal Displacements

One has to compare the regularized BIE (4.1), the regularized BCE (4.25) and the usual BIE at an internal point $p = \boldsymbol{\xi}$ (equation (1.21) in Chapter 1). The result is:

$$
u_k(\boldsymbol{\xi}) =
$$

$$
\frac{1}{2} \sum_{m=1}^{M} \sum_{\alpha=13}^{27} \left[\oint_{L_m} \left(\overline{\sigma}_{\alpha ij}(\mathbf{z}) U_{ik}(\mathbf{z}) - \overline{u}_{\alpha i}(\mathbf{z}) \Sigma_{ijk}(\mathbf{z}) \right) \epsilon_{jst} z_s dz_t \right] \left[\overset{m}{T^{-1}} \overset{m}{a} \right]_\alpha
$$

$$
+ \sum_{m=1}^{M} \sum_{\alpha=4}^{12} \left[\oint_{L_m} \left(\overline{\sigma}_{\alpha ij}(\mathbf{z}) U_{ik}(\mathbf{z}) - \overline{u}_{\alpha i}(\mathbf{z}) \Sigma_{ijk}(\mathbf{z}) \right) \epsilon_{jst} z_s dz_t \right] \left[R(\boldsymbol{\xi}) \overset{m}{T^{-1}} \overset{m}{a} \right]_{\alpha-3}
$$

$$
+ \sum_{m=1}^{M} \sum_{\alpha=1}^{3} \left[\oint_{L_m} D_{\alpha jk}(\mathbf{z}) dz_j \right] \left[S(\boldsymbol{\xi}) \overset{m}{T^{-1}} \overset{m}{a} \right]_\alpha \tag{4.48}
$$

where $\mathbf{z} = \mathbf{y} - \boldsymbol{\xi}$.

4.3.2 Displacements at Internal Points Close to the Bounding Surface

This section is concerned with evaluation of displacements at internal points that are close to the bounding surface of a body. Section 4.3.4 concerns internal stresses also at points close to the bounding surface. Please refer to Section 1.3 of this book for a similar discussion related to the BEM. (See, also, [104]).

The first step is to choose the target point $\hat{\mathbf{x}}$ at or close to the centroid of a boundary element (see Figure 1.3). Since all other terms in equation (4.48), except the solid angle, are evaluated as contour integrals, these terms are already regularized. There are at least two ways of regularizing the solid angle term in equation (4.48). The first is to evaluate the solid angle Ω (see equation (4.24)) as a line integral by employing equation (16) in Liu [87]. The second is to use a boundary contour version of equation (1.49) and still evaluate Ω as a surface integral. The latter approach is adopted in this work.

The boundary contour version of equation (1.49) can be obtained easily. This equation is:

$$u_k(\boldsymbol{\xi}) = u_k(\hat{\mathbf{x}})$$

$$+ \frac{1}{2} \sum_{m=1}^{M} \sum_{\alpha=13}^{27} \left[\oint_{L_m} \left(\overline{\sigma}_{\alpha ij}(\mathbf{z}) U_{ik}(\mathbf{z}) - \overline{u}_{\alpha i}(\mathbf{z}) \Sigma_{ijk}(\mathbf{z}) \right) \epsilon_{jst} z_s dz_t \right] \left[\overset{m}{T^{-1}} \overset{m}{a} \right]_\alpha$$

$$+ \sum_{m=1}^{M} \sum_{\alpha=4}^{12} \left[\oint_{L_m} \left(\overline{\sigma}_{\alpha ij}(\mathbf{z}) U_{ik}(\mathbf{z}) - \overline{u}_{\alpha i}(\mathbf{z}) \Sigma_{ijk}(\mathbf{z}) \right) \epsilon_{jst} z_s dz_t \right] \left[R(\boldsymbol{\xi}) \overset{m}{T^{-1}} \overset{m}{a} \right]_{\alpha-3}$$

$$+ \sum_{m=1}^{M} \sum_{\alpha=1}^{3} \left[\oint_{L_m} D_{\alpha jk}(\mathbf{z}) dz_j \right] \left[S(\boldsymbol{\xi}) \overset{m}{T^{-1}} \overset{m}{a} - S(\hat{\mathbf{x}}) \beta^{\hat{P}} \right]_\alpha \qquad (4.49)$$

where:

$$u_k(\hat{\mathbf{x}}) = \hat{\beta}_k^{\hat{P}} = \sum_{\alpha=1}^{27} S_{k\alpha}(\hat{\mathbf{x}}) \beta_\alpha^{\hat{P}} \qquad (4.50)$$

Note that the point $\hat{P}$ has coordinates $\hat{\mathbf{x}}$.

It is important to note that, on a singular element (i.e. when integration is being carried out on an element that contains the point $\hat{\mathbf{x}}$), one has:

$$\overset{m}{T^{-1}} \overset{m}{a} = \overset{m}{\beta} = \beta^{\hat{P}} \qquad (4.51)$$

In this case, the numerator of the last integrand in equation (4.49) is $\mathcal{O}(r(\boldsymbol{\xi}, \hat{\mathbf{x}}))$ while the denominator in the solid angle term is $\mathcal{O}(r^2(\boldsymbol{\xi}, \hat{\mathbf{x}}))$ as $\mathbf{y} \to \hat{\mathbf{x}}$, so that equation (4.49) is "nearly weakly singular" as $\mathbf{y} \to \hat{\mathbf{x}}$. It is useful to remember that the integral of $D_{\alpha jk}$ in equation (4.49) contains the solid angle term which is evaluated as a surface integral.

4.3.3 Internal Stresses

This time, one has to compare the regularized HBIE (4.37), the regularized HBCE (4.44) and the usual integral expression for the displacement gradient at an internal point $p = \boldsymbol{\xi}$ (equation (1.34) in Chapter 1). The result is:

$$u_{k,n}(\boldsymbol{\xi}) = - \sum_{m=1}^{M} \sum_{\alpha=13}^{27} \left[\oint_{L_m} \left(\overline{\sigma}_{\alpha ij}(\mathbf{z}) U_{ik}(\mathbf{z}) - \overline{u}_{\alpha i}(\mathbf{z}) \Sigma_{ijk}(\mathbf{z}) \right) \epsilon_{jnt} dz_t \right] \left[\overset{m}{T^{-1}} \overset{m}{a} \right]_\alpha$$

$$+ \sum_{m=1}^{M} \sum_{\alpha=4}^{12} \left[\oint_{L_m} \left(\overline{\sigma}_{\alpha ij}(\mathbf{z}) U_{ik}(\mathbf{z}) - \overline{u}_{\alpha i}(\mathbf{z}) \Sigma_{ijk}(\mathbf{z}) \right) \epsilon_{jst} z_s dz_t \right] \left[R_{,n}(\boldsymbol{\xi}) \overset{m}{T^{-1}} \overset{m}{a} \right]_{\alpha-3}$$

$$-\sum_{m=1}^{M}\sum_{\alpha=4}^{12}\left[\oint_{L_m}\left(\bar{\sigma}_{\alpha ij}(\mathbf{z})U_{ik}(\mathbf{z})-\bar{u}_{\alpha i}(\mathbf{z})\Sigma_{ijk}(\mathbf{z})\right)\epsilon_{jnt}dz_t\right]\left[R(\boldsymbol{\xi})\stackrel{m}{T^{-1}}\stackrel{m}{a}\right]_{\alpha-3}$$

$$+\sum_{m=1}^{M}\sum_{\alpha=1}^{3}\left[\oint_{L_m}D_{\alpha jk}(\mathbf{z})dz_j\right]\left[S_{,n}(\boldsymbol{\xi})\stackrel{m}{T^{-1}}\stackrel{m}{a}\right]_{\alpha}$$

$$+\sum_{m=1}^{M}\sum_{\alpha=1}^{3}\left[\oint_{L_m}\Sigma_{\alpha jk}(\mathbf{z})\epsilon_{jnt}dz_t\right]\left[S(\boldsymbol{\xi})\stackrel{m}{T^{-1}}\stackrel{m}{a}\right]_{\alpha}\qquad(4.52)$$

Hooke's law is now used to obtain the internal stress from the internal displacement gradient.

Curvatures at an internal point are given in equations (4.34 - 4.36).

4.3.4   Stresses at Internal Points Close to the Bounding Surface

As before for the case of displacement evaluation at an internal point close to the boundary of a body (see start of section 4.3.2), one has two choices with respect to the strategy for evaluation of the solid angle. Again, for the sake of uniformity, a boundary contour version of equation (1.51) is used here, together with evaluation of the solid angle as a surface integral.

A boundary contour version of equation (1.51) is obtained in a manner that is quite analogous to the approach discussed in Section 4.2.2 (see, also, [99]). The first step is to use the product rule to transform equation (1.51) to the form:

$$u_{k,n}(\boldsymbol{\xi})=u_{k,n}(\hat{\mathbf{x}})$$

$$-\int_{\partial B}\left[U_{ik}(\boldsymbol{\xi},\mathbf{y})\left[\sigma_{ij}(\mathbf{y})-\sigma_{ij}(\hat{\mathbf{x}})\right]-\Sigma_{ijk}(\boldsymbol{\xi},\mathbf{y})\left[u_i(\mathbf{y})-u_i^{(L)}\right]\right]_{,n}n_j(\mathbf{y})dS(\mathbf{y})$$

$$+\int_{\partial B}\left[U_{ik}(\boldsymbol{\xi},\mathbf{y})\sigma_{ij,n}(\mathbf{y})-\Sigma_{ijk}(\boldsymbol{\xi},\mathbf{y})\left[u_{i,n}(\mathbf{y})-u_{i,n}(\hat{\mathbf{x}})\right]\right]n_j(\mathbf{y})dS(\mathbf{y})\qquad(4.53)$$

where (see Figure 1.3):

$$u_i^{(L)}=u_i(\hat{\mathbf{x}})-u_{i,\ell}(\hat{\mathbf{x}})\hat{z}_\ell\qquad(4.54)$$

with:

$$\hat{z}_\ell=y_\ell-\hat{x}_\ell\qquad(4.55)$$

The BCM version of equation (1.51) is:

$$u_{k,n}(\boldsymbol{\xi})=u_{k,n}(\hat{\mathbf{x}})$$

$$
-\sum_{m=1}^{M}\sum_{\alpha=13}^{27}\left[\oint_{L_m}\left(\overline{\sigma}_{\alpha ij}(\hat{\mathbf{z}})U_{ik}(\mathbf{z})-\overline{u}_{\alpha i}(\hat{\mathbf{z}})\Sigma_{ijk}(\mathbf{z})\right)\epsilon_{jnt}dz_t\right]\left[\overset{m}{T}{}^{-1}\overset{m}{a}\right]_{\alpha}
$$

$$
+\sum_{m=1}^{M}\sum_{\alpha=4}^{12}\left[\oint_{L_m}\left(\overline{\sigma}_{\alpha ij}(\mathbf{z})U_{ik}(\mathbf{z})-\overline{u}_{\alpha i}(\mathbf{z})\Sigma_{ijk}(\mathbf{z})\right)\epsilon_{jst}z_s dz_t\right]\left[R_{,n}(\boldsymbol{\xi})\;\overset{m}{T}{}^{-1}\overset{m}{a}\right]_{\alpha-3}
$$

$$
-\sum_{\substack{m=1\\m\notin S}}^{M}\sum_{\alpha=4}^{12}\left[\oint_{L_m}\left(\overline{\sigma}_{\alpha ij}(\hat{\mathbf{z}})U_{ik}(\mathbf{z})-\overline{u}_{\alpha i}(\hat{\mathbf{z}})\Sigma_{ijk}(\mathbf{z})\right)\epsilon_{jnt}dz_t\right]\times
$$

$$
\left[R(\hat{\mathbf{x}})\left(\overset{m}{T}{}^{-1}\overset{m}{a}-\beta\overset{\hat{P}}{}\right)\right]_{\alpha-3}
$$

$$
+\sum_{m=1}^{M}\sum_{\alpha=1}^{3}\left[\oint_{L_m}D_{\alpha jk}(\mathbf{z})dz_j\right]\left[S_{,n}(\boldsymbol{\xi})\;\overset{m}{T}{}^{-1}\overset{m}{a}-S_{,n}(\hat{\mathbf{x}})\beta\overset{\hat{P}}{}\right]_{\alpha}
$$

$$
+\sum_{\substack{m=1\\m\notin S}}^{M}\sum_{\alpha=1}^{3}\left[\oint_{L_m}\Sigma_{\alpha jk}(\mathbf{z})\epsilon_{jnt}dz_t\right]\left[S(\hat{\mathbf{x}})\left(\overset{m}{T}{}^{-1}\overset{m}{a}-\beta\overset{\hat{P}}{}\right)\right]_{\alpha} \tag{4.56}
$$

where $\hat{\mathbf{z}} = \mathbf{y} - \hat{\mathbf{x}}$ (see Figure 1.3).

It should be noted that the first, third and fifth terms, with summations and integrals, on the right-hand side of equation (4.56), arise from the first integral in equation (4.53); while the second and fourth arise from the second integral in equation (4.53). Again, as in the case of equation (4.49), the last but one term on the right-hand side of equation (4.56) is "nearly weakly singular" ($\mathcal{O}(1/r(\boldsymbol{\xi},\hat{\mathbf{x}}))$ as $\mathbf{y} \to \hat{\mathbf{x}}$).

4.4 Numerical Results

Numerical results from the BCM, for selected 3-D examples, are available in [116], [109] and, from the HBCM, in [99]. Typical results, for a thick hollow sphere under internal pressure, are given below. For the results in Sections 4.4.1 - 4.4.3, the inner and outer radii of the sphere are 1 and 2 units, respectively, the shear modulus $G = 1$, the Poisson's ratio $\nu = 0.3$, and the internal pressure is 1. For the results in Section 4.4.4, the inner and outer radii of the sphere are 1 and 4 units, respectively, the Young's modulus $E = 1$, the Poisson's ratio $\nu = 0.25$, and the internal pressure is 1. A generic surface mesh on a one-eighth sphere is shown in Figure 4.2. Two levels of discretization - medium and fine, are used in this work. Mesh statistics are shown in Table 4.3.

4.4.1 Surface Displacements from the BCM and the HBCM

First, please note that (4.44) has two free indices, k and n, so that it represents nine equations. These equations arise from $u_{k,n}$. Different strategies are

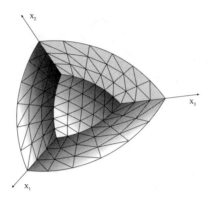

Figure 4.2: A typical mesh on the surface of a one-eighth sphere (from [100])

mesh	number of elements		
	on each flat plane	on each curved surface	total
coarse	12	9	54
medium	36	36	180
fine	64	64	320

Table 4.3: Mesh statistics on a one-eighth sphere (from [100])

possible for collocating (4.44) at a boundary point. The first is to use all nine equations. The second is to use six corresponding to $\epsilon_{kn} = (1/2)(u_{k,n} + u_{n,k})$. The six-equation strategy amounts to replacing E_{kn}, the right-hand side of (4.44), by $(1/2)(E_{kn} + E_{nk})$. Both the nine-equation and six-equation strategies lead to overdetermined systems, but are convenient for collocating at irregular boundary points since the source point normal is not involved in these cases. A third, the three-equation strategy, suitable for collocation at regular points, corresponds to the traction components τ_n. In this case, the right-hand side (E_{kn}) of (4.44) is replaced by $[\lambda E_{mm}\delta_{kn} + \mu(E_{kn} + E_{nk})]n_k(P)$, where λ and μ are Lamé constants, δ_{ij} are components of the Kronecker delta and Hooke's law is used. The three-equation strategy involving the traction components is not convenient for collocating the HBCE at a point on an edge or a corner of a body where the normal to the body surface has a jump discontinuity. In view of the assumed continuity of the stress tensor at such a point, this situation leads to a jump in traction at that point, unless the stress tensor is zero there. One would, therefore, need to use multiple source points, each belonging to a smooth surface meeting at that irregular point, and collocate separately at these points. Since the primary purpose here is to demonstrate collocation of (4.37) at irregular boundary points, only the nine-equation and six-equation strategies are used below.

It should be mentioned here that, for the HBCM in 2-D elasticity, a strategy

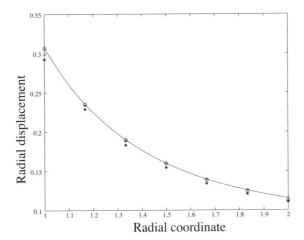

Figure 4.3: Hollow sphere under internal pressure. Radial displacement as a function of radius along the x_1 axis. Numerical solutions are obtained from the medium mesh. Exact solution: —, BCM solution: ∘ ∘ ∘∘, six-equation HBCM solution: ∗ ∗ ∗∗, nine-equation HBCM solution: ++++ (from [111])

corresponding to the first one above has been successfully employed by [131] and a strategy corresponding to the third one above has been implemented by [187].

The overdetermined system of linear algebraic equations, resulting from the nine-equation and six-equation strategies mentioned above, have been solved by using a subroutine based on QR decomposition of the system matrix. This subroutine has been obtained from the IMSL software package.

It is seen from Figure 4.2 that many of the collocation points lie on edges and six of them lie on corners of the surface of the one-eighth sphere. These results, displayed in Figure 4.3, show a comparison of the BCM (from equation (4.25)) and HBCM (from equation (4.44)) results with the exact solution of the problem [167]. The first and last points along the axis lie on corners, the rest lie along an edge. The agreement between the exact, BCM and nine-equation HBCM solution is seen to be excellent.

4.4.2 Surface Stresses

Stresses on the inner $(R = a)$ and on the outer $(R = b)$ surface of the sphere, obtained from equation (4.33) and Hooke's law, are shown in Figure 4.4. The nodes are chosen at the centroids of the boundary elements. The agreement between the numerical and analytical solutions is seen to be excellent.

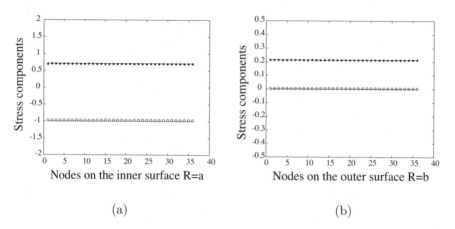

(a)   (b)

Figure 4.4: Hollow sphere under internal pressure. Stresses (a) on the inner surface $R = a$ and (b) on the outer surface $R = b$. Exact solutions —. Numerical solutions from the medium mesh: $\sigma_{\theta\theta} = \sigma_{\phi\phi}$ ****, σ_{RR} ∘∘∘∘ (from [100])

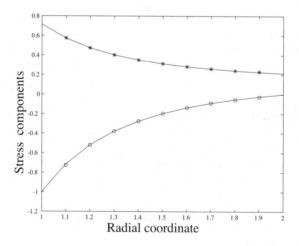

Figure 4.5: Hollow sphere under internal pressure. Stresses as functions of radius along the line $x_1 = x_2 = x_3$. Exact solutions —. Numerical solutions from the fine mesh: $\sigma_{\theta\theta} = \sigma_{\phi\phi}$ ****, σ_{RR} ∘∘∘∘ (from [103])

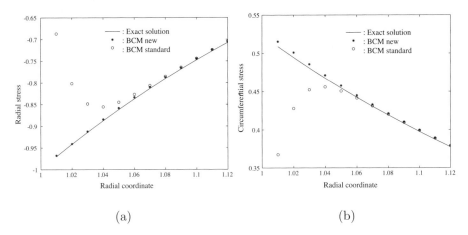

(a)   (b)

Figure 4.6: Hollow sphere under internal pressure. Radial and circumferential stresses (σ_{rr} and $\sigma_{\theta\theta}$) as functions of radial distance from the center of the sphere. The new and standard BCM solutions from the fine mesh, together with exact solutions, for points very close to the inner surface of the sphere (from [104])

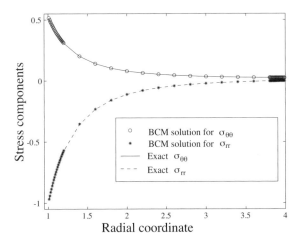

Figure 4.7: Hollow sphere under internal pressure. Radial and circumferential stresses as functions of the radial distance from the center of the sphere. Results from the new BCM (fine mesh) together with the exact solution (from [104])

4.4.3   Internal Stresses Relatively Far from the Bounding Surface

Internal stresses along the line $x_1 = x_2 = x_3$, obtained from equation (4.52) and Hooke's law, appear in Figure 4.5. Excellent agreement is observed between the numerical and analytical solutions.

4.4.4   Internal Stresses Very Close to the Bounding Surface

Numerical results for stress components, from the standard (equation (4.52)) and the new (equation (4.56)) BCM, are shown in Figures 4.6 (a) and (b), respectively. The results from the standard BCM exhibit large errors whereas results from the new BCM faithfully track the exact solutions in both cases. Finally, Figure 4.7 gives the global picture for stresses throughout the sphere. The new BCM performs beautifully, even at points that are very close to the surfaces of the hollow sphere.

Appendix A

An Explicit Form of Equation (4.40)

Using the equations :

$$u_i(\mathbf{y}) = \sum_{\alpha=1}^{27} \hat{\beta}_\alpha^m \overline{u}_{\alpha i}(z_1, z_2, z_3)$$

$$u_i(\mathbf{x}) = \sum_{\alpha=1}^{3} \hat{\beta}_\alpha^P \overline{u}_{\alpha i}(z_1, z_2, z_3)$$

$$u_{i,\ell}(\mathbf{x})[y_\ell - x_\ell] = \sum_{\alpha=4}^{12} \hat{\beta}_\alpha^P \overline{u}_{\alpha i}(z_1, z_2, z_3)$$

(the last equation above can be proved from equation (4.33)), the integral in equation (4.40) can be written as:

$$\begin{aligned}
I_1 &= \oint_{L_m} U_{ik}(\mathbf{x}, \mathbf{y}) \left[\sum_{\alpha=4}^{12} (\hat{\beta}_\alpha^m - \hat{\beta}_\alpha^P) \overline{\sigma}_{\alpha ij} + \sum_{\alpha=13}^{27} \hat{\beta}_\alpha^m \overline{\sigma}_{\alpha ij} \right] \epsilon_{jnt} dz_t \\
&\quad - \oint_{L_m} \Sigma_{ijk}(\mathbf{x}, \mathbf{y}) \left[\sum_{\alpha=1}^{12} (\hat{\beta}_\alpha^m - \hat{\beta}_\alpha^P) \overline{u}_{\alpha i} + \sum_{\alpha=13}^{27} \hat{\beta}_\alpha^m \overline{u}_{\alpha i} \right] \epsilon_{jnt} dz_t
\end{aligned}$$

where the fact that $\overline{\sigma}_{\alpha ij} = 0$ for $\alpha = 1, 2, 3$ has been used.

Appendix B

Proof of Equation (4.42)

$$u_i(\mathbf{y}) = \sum_{\alpha=1}^{27} \hat{\beta}_\alpha^m \overline{u}_{\alpha i}(z_1, z_2, z_3)$$

$$u_{i,n}(\mathbf{y}) = \sum_{\alpha=4}^{27} \hat{\beta}_\alpha^m \overline{u}_{\alpha i,n}(z_1, z_2, z_3)$$

since $\overline{u}_{\alpha i}$, $\alpha = 1, 2, 3,$ are constant.

Now,

$$u_{i,n}(\mathbf{y})_{constant} \equiv \sum_{\alpha=4}^{12} \hat{\beta}_\alpha^m \overline{u}_{\alpha i,n} = \left[S_{,N} \overset{m}{\beta} \right]_i$$

where the last equality is obtained by observing equation (4.41).

Let

$$u_{i,n}(\mathbf{y})_{linear} \equiv \sum_{\alpha=13}^{27} \hat{\beta}_\alpha^m \overline{u}_{\alpha i,n}(z_1, z_2, z_3) = \sum_{\alpha=13}^{27} \overset{m}{\beta}_\alpha \overline{u}_{\alpha i,n}(z_1, z_2, z_3)$$

Now, with n = 1,

$$\sum_{\alpha=13}^{27} \overset{m}{\beta}_\alpha \overline{\mathbf{u}}_{\alpha,1} = 2 \left(\overset{m}{\beta}_{13} + \overset{m}{\beta}_{20} \right) \overline{\mathbf{u}}_4 + 2 \overset{m}{\beta}_{15} \overline{\mathbf{u}}_5 + 2 \overset{m}{\beta}_{18} \overline{\mathbf{u}}_6$$

$$+, \ldots\ldots, + \left(k_2 \overset{m}{\beta}_{19} + k_1 \overset{m}{\beta}_{20} \right) \overline{\mathbf{u}}_{12}$$

$$= \sum_{\alpha=4}^{12} \left[R_{,1} \overset{m}{\beta} \right]_{\alpha-3} \overline{\mathbf{u}}_\alpha(z_1, z_2, z_3)$$

Similar expressions can be obtained for n = 2,3.

In the above, a vector displacement shape function is defined as:

$$\overline{\mathbf{u}}_\alpha = [\overline{u}_{\alpha 1}, \overline{u}_{\alpha 2}, \overline{u}_{\alpha 3}]^T$$

(where T denotes the transpose of the vector), and the constants k_1 and k_2 are defined in Table 4.1

Appendix C

Conversion of Equation (4.43)

$$
\overset{m}{I}_2 = -\sum_{\alpha=4}^{12}\left[\int_{S_m}\left[U_{ik}(\mathbf{x},\mathbf{y})\overline{\sigma}_{\alpha ij} - \Sigma_{ijk}(\mathbf{x},\mathbf{y})\overline{u}_{\alpha i}\right]n_j dS\right]\left[R_{,N}\,\overset{m}{\beta}\right]_{\alpha-3}
$$

$$
+\sum_{\alpha=1}^{3}\left[\int_{S_m}\Sigma_{\alpha jk}(\mathbf{x},\mathbf{y})n_j dS\right]\left[S_{,N}\,\overset{m}{\beta}\right]_{\alpha} - \sum_{\alpha=1}^{3}u_{\alpha,n}(\mathbf{x})\int_{S_m}\Sigma_{\alpha jk}(\mathbf{x},\mathbf{y})n_j dS
$$

Applying Kaplan [69]'s formula (see [116], [109]), to the first term above (its integrand is $O(1/r)$), one gets the contour integral:

$$
\overset{m}{I}_{2_1} = -\sum_{\alpha=4}^{12}\left[\oint_{L_m}(\overline{\sigma}_{\alpha ij}U_{ik} - \overline{u}_{\alpha i}\Sigma_{ijk})\,\epsilon_{jst}z_s dz_t\right]\left[R_{,N}\,\overset{m}{\beta}\right]_{\alpha-3}
$$

The remaining terms cancel on a singular element (see equation (4.41)), while, on a nonsingular element, one gets (see equation (4.26)):

$$
\overset{m}{I}_{2_2} = -\sum_{\alpha=1}^{3}\left[\oint_{L_m}D_{\alpha jk}dz_j\right]\left[S_{,N}\left(\overset{m}{\beta} - \overset{P}{\beta}\right)\right]_{\alpha}
$$

Chapter 5

SHAPE SENSITIVITY ANALYSIS

Shape sensitivity analysis of surface and internal displacements and stresses, obtained from the BCM for 3-D linear elasticity, is the subject of this chapter. Further details are available in [100, 103].

5.1 Sensitivities of Boundary Variables

The starting point in this chapter is the standard BIE (1.26) collocated at an internal point $\boldsymbol{\xi}$. (This regularized BIE is valid both at an internal point $\boldsymbol{\xi} \in B$ as well as at a boundary point $\mathbf{x} \in \partial B$.) The sensitivity (total or material derivative) of this equation is taken with respect to a design variable b. The resulting sensitivity BIE is split into three parts. The first part vanishes and the surface integrals in the second and third parts are systematically converted into contour integrals by using Stokes' theorem.

5.1.1 Sensitivity of the BIE

Equation (4.1), at an internal point $\boldsymbol{\xi} \in \partial B$ is rewritten as:

$$
\begin{aligned}
0 &= \int_{\partial B} \left[U_{ik}(\boldsymbol{\xi}, \mathbf{y}) \sigma_{ij}(\mathbf{y}, b) - \Sigma_{ijk}(\boldsymbol{\xi}, \mathbf{y})[u_i(\mathbf{y}, b) - u_i(\boldsymbol{\xi}, b)] \right] n_j(\mathbf{y}) dS(\mathbf{y}) \\
&\equiv \int_{\partial B} F_{jk}(\boldsymbol{\xi}, \mathbf{y}, b) n_j(\mathbf{y}) dS(\mathbf{y})
\end{aligned}
\tag{5.1}
$$

As mentioned above, b is a shape design variable and the spatial coordinates of the source and field points depend on b, i.e. $\boldsymbol{\xi}(b), \mathbf{y}(b)$.

From equation (5.1), one has:

$$F_{jk}(\boldsymbol{\xi},\mathbf{y},b) = U_{ik}(\boldsymbol{\xi},\mathbf{y})\sigma_{ij}(\mathbf{y},b) - \Sigma_{ijk}(\boldsymbol{\xi},\mathbf{y})[u_i(\mathbf{y},b) - u_i(\boldsymbol{\xi},b)] \qquad (5.2)$$

Now the (total) sensitivity of a function $f(\boldsymbol{\xi}(b),\mathbf{y}(b),b)$, in a materials derivative sense, is defined as :

$$\overset{*}{f} \equiv \frac{df}{db} = \frac{\partial f}{\partial \xi_k}\overset{*}{\xi}_k + \frac{\partial f}{\partial y_k}\overset{*}{y}_k + \frac{\partial f}{\partial b} \qquad (5.3)$$

while its partial sensitivity is defined as :

$$\overset{\triangle}{f} \equiv \frac{\partial f}{\partial b} \qquad (5.4)$$

It should be noted that, for $u_i(\mathbf{y},b)$:

$$\overset{\triangle}{u}_i = \overset{*}{u}_i - u_{i,k}\overset{*}{y}_k \;\; , \qquad \overset{\triangle}{\sigma}_{ij} = \overset{*}{\sigma}_{ij} - \sigma_{ij,k}\overset{*}{y}_k \qquad (5.5)$$

where , $k \equiv \frac{\partial}{\partial y_k}$; and similarly for $u_i(\boldsymbol{\xi},b)$.

Taking the sensitivity (total derivative) of equation (5.1) with respect to b, one gets:

$$
\begin{aligned}
0 \;=\; & \overset{*}{\xi}_r \int_{\partial B} \frac{\partial F_{jk}(\boldsymbol{\xi},\mathbf{y},b)}{\partial \xi_r} n_j(\mathbf{y})dS(\mathbf{y}) \\
& + \int_{\partial B} \frac{\partial F_{jk}(\boldsymbol{\xi},\mathbf{y},b)}{\partial y_r}\overset{*}{y}_r\, n_j(\mathbf{y})dS(\mathbf{y}) + \int_{\partial B} F_{jk}(\boldsymbol{\xi},\mathbf{y},b)\frac{d}{db}[n_j(\mathbf{y})dS(\mathbf{y})] \\
& + \int_{\partial B} \left[U_{ik}(\mathbf{x},\mathbf{y})\overset{\triangle}{\sigma}_{ij}(\mathbf{y},b) - \Sigma_{ijk}(\boldsymbol{\xi},\mathbf{y})(\overset{\triangle}{u}_i(\mathbf{y},b) - \overset{\triangle}{u}_i(\boldsymbol{\xi},b)) \right] n_j(\mathbf{y})dS(\mathbf{y})
\end{aligned}
$$
$$(5.6)$$

It should be noted that the last integrand above is $\frac{\partial F_{jk}(\boldsymbol{\xi},\mathbf{y},b)}{\partial b}$.

The first integral on the right-hand side of equation (5.6) is zero because the integral in equation (5.1) vanishes for all values of $\boldsymbol{\xi} \in B$. Let the second and third integrals together be called I_k and the last integral J_k. Thus:

$$I_k + J_k = 0 \qquad (5.7)$$

Each of these surface integrals will be converted to line integrals in the next two sections.

5.1.2   The Integral I_k

The surface integral I_k is converted to a sum of contour integrals in this section. As mentioned before, a point $\boldsymbol{\xi}$, inside the body, is considered first. From Bonnet and Xiao [15] (see also Petryk and Mróz [128]):

$$\overset{*}{n}_j = -\overset{*}{y}_{r,j}\, n_r + \overset{*}{y}_{r,m}\, n_r n_m n_j \qquad (5.8)$$

$$\frac{d\overset{*}{S}}{dS} = \overset{*}{y}_{r,r} - \overset{*}{y}_{r,m}\, n_r n_m \tag{5.9}$$

so that:

$$\frac{d}{db}\left[n_j dS\right] = \left[\overset{*}{y}_{r,r}\, n_j - \overset{*}{y}_{r,j}\, n_r\right] dS \tag{5.10}$$

Using equation (5.10) in (5.6), one has:

$$I_k = \int_{\partial B}\left[F_{jk,r}\,\overset{*}{y}_r\, n_j + F_{jk}\,\overset{*}{y}_{r,r}\, n_j - F_{jk}\,\overset{*}{y}_{r,j}\, n_r\right] dS \tag{5.11}$$

Since $F_{jk,j} = 0$ (except at a point of singularity - here ξ is an internal point), the above expression can be written as:

$$I_k = \int_{\partial B}\left[\left(F_{jk}\,\overset{*}{y}_r\right)_{,r} n_j - \left(F_{jk}\,\overset{*}{y}_r\right)_{,j} n_r\right] dS \tag{5.12}$$

Let $\partial B = \cup S_m$ and let L_m be the bounding contour of the boundary element S_m. Using Stokes' theorem in the form (see Toh and Mukherjee [168]):

$$\int_{S_m}\left(F_{,r} n_j - F_{,j} n_r\right) dS(\mathbf{y}) = \oint_{L_m} \epsilon_{qjr} F\, dy_q \tag{5.13}$$

with $F \equiv F_{jk}\,\overset{*}{y}_r$, one gets:

$$I_k = \sum_{m=1}^{M} \oint_{L_m} \epsilon_{qjr} F_{jk}\,\overset{*}{y}_r\, dy_q \tag{5.14}$$

It is useful to state here that formula (5.14) is a general one, in the sense that, a surface integral I (over a closed surface ∂B) of any divergence-free vector function $\mathbf{F}(\mathbf{y})$, of the form:

$$I = \int_{\partial B} F_j(\mathbf{y})\, n_j(\mathbf{y})\, dS(\mathbf{y}) \tag{5.15}$$

has the sensitivity expression :

$$\overset{*}{I} = \sum_{m=1}^{M} \oint_{L_m} \epsilon_{jnt} F_j(\mathbf{y})\,\overset{*}{y}_n\, dy_t \tag{5.16}$$

Of course, from Gauss' theorem, $I = 0$. Therefore, $\overset{*}{I} = 0$.
One can show that:

$$\sum_{m=1}^{M} \oint_{L_m} \epsilon_{qjr} F_{jk} dy_q = \int_{\partial B} F_{jk,r} n_j dS = 0 \;\;\Rightarrow\;\; \sum_{m=1}^{M} \overset{*}{\xi}_r \oint_{L_m} \epsilon_{qjr} F_{jk} dy_q = 0 \tag{5.17}$$

Since $\overset{*}{y}_r = \overset{*}{\xi}_r + \overset{*}{z}_r$, one can replace $\overset{*}{y}_r$ with $\overset{*}{z}_r$ in equation (5.14). Also, $dy_q = dz_q$ since $dx_q = 0$ at a fixed source point.

Substituting equation (5.2) into (5.14) (with $\mathbf{y}$ replaced by $\mathbf{z}$) , one gets:

$$I_k(\boldsymbol{\xi}) = \sum_{m=1}^{M} \oint_{L_m} [U_{ik}(\boldsymbol{\xi},\mathbf{y})\sigma_{ij}(\mathbf{y},b) - \Sigma_{ijk}(\boldsymbol{\xi},\mathbf{y})(u_i(\mathbf{y},b) - u_i(\boldsymbol{\xi},b))] \,\epsilon_{jnt} \overset{*}{z}_n \, dz_t$$

(5.18)

Since $\overset{*}{z}_n$ is $O(r)$ as $r = \| \mathbf{y} - \mathbf{x} \| \to 0$, the above expression is completely regular. Therefore, it remains valid at a boundary point $\mathbf{x} \in \partial B$.

5.1.3   The Integral J_k

The surface integral J_k is converted to a sum of contour integrals in this section.

5.1.3.1   Shape functions for partial sensitivities - a simple example

Series expressions for the partial sensitivities of displacements and stresses, in terms of global BCM shape functions, are derived next.

It is useful to start with a very simple example. Let:

$$f(y) = a_0 + a_1 y + a_2 y^2$$

(5.19)

With the change of variables:

$$y = x + z$$

(5.20)

one can write:

$$f(x, z) = \hat{a}_0(x) + \hat{a}_1(x)z + \hat{a}_2(x)z^2$$

(5.21)

where:

$$
\begin{aligned}
\hat{a}_0 &= a_0 + a_1 x + a_2 x^2 = f(x) \\
\hat{a}_1 &= a_1 + 2a_2 x = \frac{d}{dx} f(x) \\
\hat{a}_2 &= a_2
\end{aligned}
$$

(5.22)

It is easy to show that taking sensitivities of the above equations results in:

$$\overset{*}{f}(y) = \overset{*}{\hat{a}}_0 + \overset{*}{\hat{a}}_1 y + \overset{*}{\hat{a}}_2 y^2 + a_1 \overset{*}{y} + 2a_2 y \overset{*}{y}$$

(5.23)

$$
\begin{aligned}
\overset{\triangle}{f}(y) &= \overset{*}{f}(y) - f_{,y} \overset{*}{y} \\
&= \overset{*}{\hat{a}}_0 + \overset{*}{\hat{a}}_1 y + \overset{*}{\hat{a}}_2 y^2
\end{aligned}
$$

(5.24)

As expected, the partial sensitivity of $f(x)$ is its sensitivity at a fixed point in space.

Now, with the change of variables (5.20), one has:

$$\overset{\triangle}{f}(x, z) = \overset{\diamond}{a}_0(x) + \overset{\diamond}{a}_1(x)z + \overset{\diamond}{a}_2(x)z^2 \tag{5.25}$$

where $\overset{\diamond}{a}_k$, $k = 0, 1, 2$, are related to $\overset{\diamond}{a}_k$ in the same manner as $\hat{a}_k$ are related to a_k in equation (5.22).

5.1.3.2   BCM interpolation functions

The displacement and stress interpolation functions for the BCM are considered next.

One starts with equation (4.8) for the displacements. Following the procedure outlined above for the simple example, it is easy to show that:

$$\overset{\triangle}{u}_i = \sum_{\alpha=1}^{27} \overset{*}{\beta}_\alpha \bar{u}_{\alpha i}(y_1, y_2, y_3) = \sum_{\alpha=1}^{27} \overset{\diamond}{\beta}_\alpha (x_1, x_2, x_3)\bar{u}_{\alpha i}(z_1, z_2, z_3) \tag{5.26}$$

where, $\overset{\diamond}{\beta}_\alpha, \alpha = 1, 2, ...27$, are related to $\overset{*}{\beta}_\alpha$ in the same manner as $\hat{\beta}_\alpha$ are related to β_α (see equations 4.10, 4.11 and 4.12), i.e.:

$$\overset{\diamond}{\beta}_i = \sum_{\alpha=1}^{27} S_{i\alpha}(x_1, x_2, x_3)\overset{*}{\beta}_\alpha \ , \quad i = 1, 2, 3 \tag{5.27}$$

$$\overset{\diamond}{\beta}_k = \sum_{\alpha=1}^{27} R_{n\alpha}(x_1, x_2, x_3)\overset{*}{\beta}_\alpha \ , \quad k = 4, 5, ...12, \quad n = k - 3 \tag{5.28}$$

$$\overset{\diamond}{\beta}_\alpha = \overset{*}{\beta}_\alpha \ , \quad\quad\quad \alpha = 13, 14, .., 27 \tag{5.29}$$

Of course, the partial sensitivities for the stresses can now be expressed as:

$$\overset{\triangle}{\sigma}_{ij} = \sum_{\alpha=1}^{27} \overset{*}{\beta}_\alpha \bar{\sigma}_{\alpha ij}(y_1, y_2, y_3) = \sum_{\alpha=1}^{27} \overset{\diamond}{\beta}_\alpha (x_1, x_2, x_3)\bar{\sigma}_{\alpha ij}(z_1, z_2, z_3) \tag{5.30}$$

5.1.3.3   The final form of J_k

The conversion procedure is entirely analogous to the derivation of the primary BCM equation presented in Chapter 4. Series expansions (5.26) and (5.30) are first substituted into the last integral on the right-hand side of the sensitivity BIE (5.6). The potential functions are the same as before. Finally, the surface integral J_k is converted into a sum of contour integrals. The result is:

$$J_k(\mathbf{x}) = \frac{1}{2}\sum_{m=1}^{M}\sum_{\alpha=13}^{27}\left[\oint_{L_m}(\overline{\sigma}_{\alpha ij}U_{ik}-\overline{u}_{\alpha i}\Sigma_{ijk})\,\epsilon_{jnt}z_n dz_t\right]\overset{\diamond m}{\beta_\alpha}$$

$$+\sum_{m=1}^{M}\sum_{\alpha=4}^{12}\left[\oint_{L_m}(\overline{\sigma}_{\alpha ij}U_{ik}-\overline{u}_{\alpha i}\Sigma_{ijk})\,\epsilon_{jnt}z_n dz_t\right]\overset{\diamond m}{\beta_\alpha}$$

$$+\sum_{\substack{m=1\\m\notin S}}^{M}\sum_{\alpha=1}^{3}\left[\oint_{L_m}D_{\alpha jk}dz_j\right]\left[\overset{\diamond m}{\beta_\alpha}-\overset{\diamond P}{\beta_\alpha}\right] \qquad (5.31)$$

It is interesting to comment on the physical meaning of the quantities $\overset{\diamond P}{\beta_\alpha}$ for the case when the surface source point P is a regular off-contour boundary point (ROCBP). Comparing equations (4.10 - 4.12) with (5.27 - 5.29), it is clear that $\overset{\triangle}{\hat{\beta}_k}=\overset{\diamond}{\hat{\beta}_k}$, $k = 1, 2, 3, .., 27$. Now, the quantities $\overset{\diamond P}{\beta_\alpha}$ can be easily interpreted in terms of the partial sensitivities of displacements and their derivatives from equations (4.32 - 4.36). For example, $\overset{\diamond P}{\beta_k}=\overset{\triangle}{\hat{u}_k}(P)$, $k = 1, 2, 3$, etc.

5.1.4 The BCM Sensitivity Equation

An explicit form of the BCM sensitivity equation is now derived.

On any boundary element:

$$\overset{*}{\beta}= T^{-1}\overset{*}{a}+(T^{-1})^*a \qquad (5.32)$$

in which it is convenient to evaluate the sensitivity of T^{-1} from the formula:

$$(T^{-1})^* = -T^{-1}\overset{*}{T}T^{-1} \qquad (5.33)$$

Expressions (5.27 - 5.29) for $\overset{\diamond}{\beta_\alpha}$ are substituted into (5.31), and (5.32) is used to write $\overset{*}{\beta}$ in explicit form. Next, an explicit expression for I_k is obtained by substituting the series expressions (4.8) for u_i and (4.9) for σ_{ij} into (5.18). Finally, the explicit expression for I_k (obtained from (5.18)) and the explicit expression for J_k (obtained from (5.31)) are input into the BIE sensitivity equation (5.6) (see also 5.7) evaluated at a general boundary point $\mathbf{x} \in \partial B$. The result is:

$$0 = \sum_{m=1}^{M}\sum_{\alpha=13}^{27}\left[\oint_{L_m}(\overline{\sigma}_{\alpha ij}U_{ik}-\overline{u}_{\alpha i}\Sigma_{ijk})\,\epsilon_{jnt}\overset{*}{z}_n\,dz_t\right]\left[\overset{m}{T^{-1}\overset{m}{a}}\right]_\alpha$$

$$+\sum_{m=1}^{M}\sum_{\alpha=4}^{12}\left[\oint_{L_m}(\overline{\sigma}_{\alpha ij}U_{ik}-\overline{u}_{\alpha i}\Sigma_{ijk})\,\epsilon_{jnt}\overset{*}{z}_n\,dz_t\right]\left[R\,\overset{m}{T^{-1}\overset{m}{a}}\right]_{\alpha-3}$$

$$-\sum_{\substack{m=1\\m\notin S}}^{M}\sum_{\alpha=1}^{3}\left[\oint_{L_m}\Sigma_{\alpha jk}\epsilon_{jnt}\overset{*}{z}_n\,dz_t\right]\left[S\left(\overset{m}{T}{}^{-1}\overset{m}{a}-\overset{P}{T}{}^{-1}\overset{P}{a}\right)\right]_\alpha$$

$$+\frac{1}{2}\sum_{m=1}^{M}\sum_{\alpha=13}^{27}\left[\oint_{L_m}(\overline{\sigma}_{\alpha ij}U_{ik}-\overline{u}_{\alpha i}\Sigma_{ijk})\,\epsilon_{jnt}z_n dz_t\right]\times$$
$$\left[\overset{m}{T}{}^{-1}\overset{*m}{a}+\left(\overset{m}{T}{}^{-1}\right)^{*}\overset{m}{a}\right]_\alpha$$

$$+\sum_{m=1}^{M}\sum_{\alpha=4}^{12}\left[\oint_{L_m}(\overline{\sigma}_{\alpha ij}U_{ik}-\overline{u}_{\alpha i}\Sigma_{ijk})\,\epsilon_{jnt}z_n dz_t\right]\times$$
$$\left[R\left(\overset{m}{T}{}^{-1}\overset{*m}{a}+\left(\overset{m}{T}{}^{-1}\right)^{*}\overset{m}{a}\right)\right]_{\alpha-3}$$

$$+\sum_{\substack{m=1\\m\notin S}}^{M}\sum_{\alpha=1}^{3}\left[\oint_{L_m}D_{\alpha jk}dz_j\right]\times$$
$$\left[S\left(\overset{m}{T}{}^{-1}\overset{*m}{a}-\overset{P}{T}{}^{-1}\overset{*P}{a}+\left(\overset{m}{T}{}^{-1}\right)^{*}\overset{m}{a}-\left(\overset{P}{T}{}^{-1}\right)^{*}\overset{P}{a}\right)\right]_\alpha$$

$$(5.34)$$

Comparison of the above sensitivity equation (5.34) with the standard BCE (4.25) reveals that the integrals in its last three terms are identical to those in the standard BCE. Therefore, its discretized form can be written with the same coefficient matrix A as for the standard BCM, i.e. :

$$K\overset{*}{a}=h \tag{5.35}$$

where the right-hand side vector $h = -\overset{*}{K}a$ can be computed from equation (5.34) by using the boundary values of the primary variables a that are known at this stage.

Finally, the usual switching of columns leads to:

$$A\overset{*}{x}=B\overset{*}{y}+h \tag{5.36}$$

where $\overset{*}{x}$ contains the unknown and $\overset{*}{y}$ the known values of boundary sensitivities. In many applications, $\overset{*}{y}=0$

5.2   Sensitivities of Surface Stresses

The first step is to use equation (5.32) to find $\overset{*}{\beta}$ on each element.

There are at least four ways to find stress sensitivities on the surface of a body.

5.2.1 Method One

Equation (4.9) is differentiated to give:

$$\overset{*}{\sigma}_{ij} = \sum_{\alpha=1}^{27} \overset{*}{\beta}_\alpha \, \overline{\sigma}_{\alpha ij}(y_1, y_2, y_3) + \sum_{\alpha=1}^{27} \beta_\alpha \, \overset{*}{\overline{\sigma}}_{\alpha ij} \, (\overset{*}{y}_1, \overset{*}{y}_2, \overset{*}{y}_3) \qquad (5.37)$$

Note that $\overset{*}{\overline{\sigma}}_{\alpha ij} \, (\overset{*}{y}_1, \overset{*}{y}_2, \overset{*}{y}_3)$ involves sensitivities of the field point coordinates (y_1, y_2, y_3).

5.2.2 Method Two

Sensitivities of displacement gradients $v_{ij} \equiv u_{i,j}$ are computed by differentiating equation (4.33). The result is:

$$\left[\overset{*}{v}_{ij} (\mathbf{x}) \right] = \begin{bmatrix} \overset{*}{\hat{\beta}}_4 & \overset{*}{\hat{\beta}}_7 & \overset{*}{\hat{\beta}}_{10} \\ \overset{*}{\hat{\beta}}_5 & \overset{*}{\hat{\beta}}_8 & \overset{*}{\hat{\beta}}_{11} \\ \overset{*}{\hat{\beta}}_6 & \overset{*}{\hat{\beta}}_9 & \overset{*}{\hat{\beta}}_{12} \end{bmatrix}_{\mathbf{x}} \qquad (5.38)$$

where, by differentiating equation (4.11), one has:

$$\overset{*}{\hat{\beta}}_k = \sum_{\alpha=1}^{27} R_{n\alpha}(x_1, x_2, x_3) \, \overset{*}{\beta}_\alpha + \sum_{\alpha=1}^{27} \overset{*}{R}_{n\alpha} \, (\overset{*}{x}_1, \overset{*}{x}_2, \overset{*}{x}_3)\beta_\alpha$$

$$k = 4, 5, ...12, \quad n = k - 3 \qquad (5.39)$$

Note that $\overset{*}{R}_{n\alpha} \, (\overset{*}{x}_1, \overset{*}{x}_2, \overset{*}{x}_3)$ involves sensitivities of the source point coordinates (x_1, x_2, x_3)

Finally, Hooke's law is used to determine the stress sensitivities from the sensitivities of the displacement gradients.

5.2.3 Method Three

One writes:

$$\overset{*}{v}_{ij} = \overset{\triangle}{v}_{ij} + v_{ij,k} \, \overset{*}{x}_k \qquad (5.40)$$

Now:

$$v_{ij,k} = u_{i,jk} \qquad (5.41)$$

Also, from equation (4.33):

$$\left[\overset{\triangle}{v}_{ij}(\mathbf{x}) \right] = \begin{bmatrix} \overset{\triangle}{\hat{\beta}}_4 & \overset{\triangle}{\hat{\beta}}_7 & \overset{\triangle}{\hat{\beta}}_{10} \\ \overset{\triangle}{\hat{\beta}}_5 & \overset{\triangle}{\hat{\beta}}_8 & \overset{\triangle}{\hat{\beta}}_{11} \\ \overset{\triangle}{\hat{\beta}}_6 & \overset{\triangle}{\hat{\beta}}_9 & \overset{\triangle}{\hat{\beta}}_{12} \end{bmatrix}_{\mathbf{x}} \tag{5.42}$$

with, from equations (4.11) and (5.28):

$$\overset{\triangle}{\hat{\beta}}_k = \overset{\diamond}{\hat{\beta}}_k = \sum_{\alpha=1}^{27} R_{n\alpha}(x_1, x_2, x_3) \overset{*}{\beta}_\alpha$$

$$k = 4, 5, ..., 12, \quad n = k - 3 \tag{5.43}$$

The curvature expressions needed in equation (5.40) are available in terms of β_α, $\alpha = 13, 14, ...27$, from equations (4.34 - 4.36).

5.2.4 Method Four

The starting point is, again, equation (5.40). The term $v_{ij,k}$ on the right-hand side of equation (5.40) is treated in the same fashion as in Section 5.2.3 . The other term, $\overset{\triangle}{v}_{ij}$, is treated differently. It is first observed that the operators , k and $\triangle$ commute and that $\overset{\triangle}{u}_i(P) = \overset{\diamond P}{\beta}_i$ for $i = 1, 2, 3$. Therefore, one has:

$$\overset{\triangle}{v}_{ij} = (u_{i,j})^{\triangle} = (\overset{\triangle}{u}_i)_{,j} = (\overset{\diamond P}{\beta}_i)_{,j} \tag{5.44}$$

It follows from equation (5.27) that:

$$\overset{\triangle}{v}_{ij} = (\overset{\diamond P}{\beta}_i)_{,j} = \sum_{\alpha=1}^{27} S_{i\alpha,j}(x_1, x_2, x_3) \overset{*}{\beta}_\alpha, \quad i, j = 1, 2, 3 \tag{5.45}$$

5.3 Sensitivities of Variables at Internal Points

Boundary contour integral equations of the sensitivities of internal displacements and stresses are derived in this section. Further details are available [103].

5.3.1 Sensitivities of Displacements

The starting point is equation (1.21), the displacement BIE at an internal point ξ. This equation is written as:

$$u_k(\boldsymbol{\xi}, b) = \int_{\partial B} \left[U_{ik}(\boldsymbol{\xi}, \mathbf{y})\sigma_{ij}(\mathbf{y}, b) - \Sigma_{ijk}(\boldsymbol{\xi}, \mathbf{y})u_i(\mathbf{y}, b)\right] n_j(\mathbf{y})dS(\mathbf{y}) \qquad (5.46)$$

Define:

$$G_{jk}(\boldsymbol{\xi}, \mathbf{y}, b) = U_{ik}(\boldsymbol{\xi}, \mathbf{y})\sigma_{ij}(\mathbf{y}, b) - \Sigma_{ijk}(\boldsymbol{\xi}, \mathbf{y})u_i(\mathbf{y}, b) \qquad (5.47)$$

Taking the sensitivity of equation (5.46), one gets:

$$
\begin{aligned}
\overset{*}{u}_k (\boldsymbol{\xi}, b) &= \overset{*}{\xi}_r \int_{\partial B} \frac{\partial G_{jk}(\boldsymbol{\xi}, \mathbf{y}, b)}{\partial \xi_r} n_j(\mathbf{y})dS(\mathbf{y}) \\
&+ \int_{\partial B} \frac{\partial G_{jk}(\boldsymbol{\xi}, \mathbf{y}, b)}{\partial y_r} \overset{*}{y}_r \, n_j(\mathbf{y})dS(\mathbf{y}) \\
&+ \int_{\partial B} G_{jk}(\boldsymbol{\xi}, \mathbf{y}, b) \frac{d}{db} \left[n_j(\mathbf{y})dS(\mathbf{y}) \right] \\
&+ \int_{\partial B} \left[U_{ik}(\boldsymbol{\xi}, \mathbf{y}) \overset{*}{\sigma}_{ij} (\mathbf{y}, b) - \Sigma_{ijk}(\boldsymbol{\xi}, \mathbf{y}) \overset{*}{u}_i (\mathbf{y}, b) \right] n_j(\mathbf{y})dS(\mathbf{y})
\end{aligned}
$$

$$(5.48)$$

Let the first term on the right-hand side of equation (5.48) be called J_{1k} , the second and third integrals together be called J_{2k} and the last integral be J_{3k}. Therefore, one has:

$$\overset{*}{u}_k (\boldsymbol{\xi}, b) = J_{1k} + J_{2k} + J_{3k} \qquad (5.49)$$

It is obvious that:

$$J_{1k} = u_{k,r}(\boldsymbol{\xi}) \overset{*}{\xi}_r \qquad (5.50)$$

Using exactly the same procedure described in Section 5.1.2, one gets:

$$J_{2k} = \sum_{m=1}^{M} \oint_{L_m} \epsilon_{jnt} G_{jk} \overset{*}{z}_n \, dz_t \qquad (5.51)$$

Finally,

$$J_{3k} = \overset{\triangle}{u}_k \qquad (5.52)$$

Substituting equations (5.50, 5.51, 5.52 and 5.47) into (5.49), one obtains the expression:

$$
\begin{aligned}
\overset{*}{u}_k (\boldsymbol{\xi}, b) &= u_{k,p}(\boldsymbol{\xi}) \overset{*}{\xi}_p \\
&+ \sum_{m=1}^{M} \oint_{L_m} \left[U_{ik}(\boldsymbol{\xi}, \mathbf{y})\sigma_{ij}(\mathbf{y}, b) - \Sigma_{ijk}(\boldsymbol{\xi}, \mathbf{y})u_i(\mathbf{y}, b)\right] \epsilon_{jnt} \overset{*}{z}_n \, dz_t + \overset{\triangle}{u}_k
\end{aligned}
$$

$$(5.53)$$

An explicit form of equation (5.53) is obtained by writing the displacements and stresses in terms of their interpolation functions from equations (4.8 and 4.9). Please refer to Section 5.1.3 for details of the treatment of the partial sensitivity $\overset{\triangle}{u}_k$. Finally, the boundary contour integral form of the displacement sensitivity equation is:

$$
\begin{aligned}
\overset{*}{u}_k(\boldsymbol{\xi}, b) \;=\;& u_{k,r}(\boldsymbol{\xi})\,\overset{*}{\xi}_r \\
&+ \sum_{m=1}^{M}\sum_{\alpha=13}^{27}\left[\oint_{L_m}(\overline{\sigma}_{\alpha ij}U_{ik}-\overline{u}_{\alpha i}\Sigma_{ijk})\,\epsilon_{jnt}\,\overset{*}{z}_n\,dz_t\right]\left[\overset{m}{T^{-1}}\overset{m}{a}\right]_\alpha \\
&+ \sum_{m=1}^{M}\sum_{\alpha=4}^{12}\left[\oint_{L_m}(\overline{\sigma}_{\alpha ij}U_{ik}-\overline{u}_{\alpha i}\Sigma_{ijk})\,\epsilon_{jnt}\,\overset{*}{z}_n\,dz_t\right]\left[R\,\overset{m}{T^{-1}}\overset{m}{a}\right]_{\alpha-3} \\
&- \sum_{m=1}^{M}\sum_{\alpha=1}^{3}\left[\oint_{L_m}\Sigma_{\alpha jk}\epsilon_{jnt}\,\overset{*}{z}_n\,dz_t\right]\left[S\,\overset{m}{T^{-1}}\overset{m}{a}\right]_\alpha \\
&+ \frac{1}{2}\sum_{m=1}^{M}\sum_{\alpha=13}^{27}\left[\oint_{L_m}(\overline{\sigma}_{\alpha ij}U_{ik}-\overline{u}_{\alpha i}\Sigma_{ijk})\,\epsilon_{jnt}z_n dz_t\right]\times \\
&\qquad\left[\overset{m}{T^{-1}}\overset{*}{\overset{m}{a}}+\left(\overset{m}{T^{-1}}\right)^{*}\overset{m}{a}\right]_\alpha \\
&+ \sum_{m=1}^{M}\sum_{\alpha=4}^{12}\left[\oint_{L_m}(\overline{\sigma}_{\alpha ij}U_{ik}-\overline{u}_{\alpha i}\Sigma_{ijk})\,\epsilon_{jnt}z_n dz_t\right]\times \\
&\qquad\left[R\left(\overset{m}{T^{-1}}\overset{*}{\overset{m}{a}}+\left(\overset{m}{T^{-1}}\right)^{*}\overset{m}{a}\right)\right]_{\alpha-3} \\
&+ \sum_{m=1}^{M}\sum_{\alpha=1}^{3}\left[\oint_{L_m}D_{\alpha jk}dz_j\right]\left[S\left(\overset{m}{T^{-1}}\overset{*}{\overset{m}{a}}+\left(\overset{m}{T^{-1}}\right)^{*}\overset{m}{a}\right)\right]_\alpha
\end{aligned}
$$

$$(5.54)$$

5.3.2 Sensitivities of Displacement Gradients and Stresses

This time, the starting point is the displacement gradient BIE (1.28), which is written as:

$$
\begin{aligned}
v_{kr}(\boldsymbol{\xi}, b) \;\equiv\;& u_{k,r}(\boldsymbol{\xi}, b) \\
=\;& -\int_{\partial B}\left[U_{ik,r}(\boldsymbol{\xi}, \mathbf{y})\sigma_{ij}(\mathbf{y}, b)-\Sigma_{ijk,r}(\boldsymbol{\xi}, \mathbf{y})u_i(\mathbf{y}, b)\right]n_j(\mathbf{y})dS(\mathbf{y})
\end{aligned}
$$

$$(5.55)$$

Now, one defines:

$$H_{jkr}(\boldsymbol{\xi}, \mathbf{y}) = U_{ik,r}(\boldsymbol{\xi}, \mathbf{y})\sigma_{ij}(\mathbf{y}, b) - \Sigma_{ijk,r}(\boldsymbol{\xi}, \mathbf{y})u_i(\mathbf{y}, b) \qquad (5.56)$$

One has $H_{jkr,j} = 0$ at an internal point since:

$$H_{jkr}(\boldsymbol{\xi}, \mathbf{y}) = -\frac{\partial G_{jk}(\boldsymbol{\xi}, \mathbf{y})}{\partial \xi_r} \qquad (5.57)$$

and $G_{jk,j} = 0$.

The exact same reasoning as in the previous section 5.3.1 leads to an equation for the sensitivities of displacement gradients at an internal point. This is:

$$
\begin{aligned}
\overset{*}{v}_{kr}(\boldsymbol{\xi}, b) = {} & u_{k,rp}(\boldsymbol{\xi})\, \overset{*}{\xi}_p \\
& - \sum_{m=1}^{M} \oint_{L_m} [U_{ik,r}(\boldsymbol{\xi}, \mathbf{y})\sigma_{ij}(\mathbf{y}, b) - \Sigma_{ijk,r}(\boldsymbol{\xi}, \mathbf{y})u_i(\mathbf{y}, b)]\, \epsilon_{jnt}\, \overset{*}{z}_n\, dz_t \\
& - \int_{\partial B} \left[U_{ik,r}(\boldsymbol{\xi}, \mathbf{y})\, \overset{\triangle}{\sigma}_{ij}(\mathbf{y}, b) - \Sigma_{ijk,r}(\boldsymbol{\xi}, \mathbf{y})\, \overset{\triangle}{u}_i(\mathbf{y}, b) \right] n_j(\mathbf{y})dS(\mathbf{y})
\end{aligned}
$$
$$(5.58)$$

It is very interesting to verify (5.58) from another point of view. Differentiating equation (5.53) with respect to ξ_r, one gets:

$$
\begin{aligned}
(\overset{*}{u}_k)_{,r}(\boldsymbol{\xi}) = {} & u_{k,rp}(\boldsymbol{\xi})\, \overset{*}{\xi}_p + u_{k,p}(\boldsymbol{\xi})(\overset{*}{\xi}_p)_{,r} \\
& - \sum_{m=1}^{M} \oint_{L_m} [U_{ik,r}(\boldsymbol{\xi}, \mathbf{y})\sigma_{ij}(\mathbf{y}, b) - \Sigma_{ijk,r}(\boldsymbol{\xi}, \mathbf{y})u_i(\mathbf{y}, b)]\, \epsilon_{jnt}\, \overset{*}{z}_n\, dz_t \\
& - \int_{\partial B} \left[U_{ik,r}(\boldsymbol{\xi}, \mathbf{y})\, \overset{\triangle}{\sigma}_{ij}(\mathbf{y}, b) - \Sigma_{ijk,r}(\boldsymbol{\xi}, \mathbf{y})\, \overset{\triangle}{u}_i(\mathbf{y}, b) \right] n_j(\mathbf{y})dS(\mathbf{y})
\end{aligned}
$$
$$(5.59)$$

Normally, one would expect another term on the right-hand side of the above equation, namely:

$$\sum_{m=1}^{M} \oint_{L_m} [U_{ik}(\boldsymbol{\xi}, \mathbf{y})\sigma_{ij}(\mathbf{y}, b) - \Sigma_{ijk}(\boldsymbol{\xi}, \mathbf{y})u_i(\mathbf{y}, b)]\, \epsilon_{jnt}(\overset{*}{z}_n)_{,r}dz_t \qquad (5.60)$$

However, (see Section 5.1.2) $\overset{*}{z}_n = \overset{*}{y}_n - \overset{*}{\xi}_n$, $(\overset{*}{z}_n)_{,r} = -(\overset{*}{\xi}_n)_{,r}$ and:

$$\sum_{m=1}^{M} (\overset{*}{\xi}_n)_{,r} \oint_{L_m} G_{jk}(\boldsymbol{\xi}, \mathbf{y}, b)\epsilon_{jnt}dz_t = 0 \qquad (5.61)$$

since G_{jk} is divergence free (see (5.17)). Therefore, the expression in (5.60) vanishes.

Now, using the well-known formula (which is valid for any sufficiently smooth function ϕ - see, for example, Haug et al. [63]):

$$(\phi_{,r})^* = (\overset{*}{\phi})_{,r} - \phi_{,p}(\overset{*}{x}_p)_{,r} \tag{5.62}$$

with $\phi = u_k$, it is easy to show that equations (5.58) and (5.59) are consistent.

Finally, one writes the displacements and stresses in terms of their interpolation functions in order to obtain an explicit form of equation (5.58). It should be noted that the second term in the right-hand side of equation (5.58) is analogous to the integral on the right-hand side of equation (5.53) (with **G** replaced by **H**), while the last term in equation (5.58), $\overset{\triangle}{v}_{kr}$, is analogous to an expression for the displacement gradient at an internal point. The displacement gradient BCE (4.52) is very useful for obtaining an explicit expression for this integral. The final result is:

$$
\begin{aligned}
\overset{*}{v}_{kr}(\boldsymbol{\xi}, b) = & \; u_{k,rp}(\boldsymbol{\xi})\,\overset{*}{\xi}_p \\
& - \sum_{m=1}^{M}\sum_{\alpha=13}^{27}\left[\oint_{L_m}(\overline{\sigma}_{\alpha ij}U_{ik,r} - \overline{u}_{\alpha i}\Sigma_{ijk,r})\,\epsilon_{jnt}\,\overset{*}{z}_n\,dz_t\right]\left[\overset{m}{T}^{-1}\overset{m}{a}\right]_\alpha \\
& - \sum_{m=1}^{M}\sum_{\alpha=4}^{12}\left[\oint_{L_m}(\overline{\sigma}_{\alpha ij}U_{ik,r} - \overline{u}_{\alpha i}\Sigma_{ijk,r})\,\epsilon_{jnt}\,\overset{*}{z}_n\,dz_t\right] \times \\
& \quad \left[R\,\overset{m}{T}^{-1}\overset{m}{a}\right]_{\alpha-3} \\
& + \sum_{m=1}^{M}\sum_{\alpha=1}^{3}\left[\oint_{L_m}\Sigma_{\alpha jk,r}\epsilon_{jnt}\,\overset{*}{z}_n\,dz_t\right]\left[S\,\overset{m}{T}^{-1}\overset{m}{a}\right]_\alpha \\
& - \sum_{m=1}^{M}\sum_{\alpha=13}^{27}\left[\oint_{L_m}(\overline{\sigma}_{\alpha ij}U_{ik} - \overline{u}_{\alpha i}\Sigma_{ijk})\,\epsilon_{jrt}dz_t\right] \times \\
& \quad \left[\overset{m}{T}^{-1}\overset{*m}{a} + \left(\overset{m}{T}^{-1}\right)^*\overset{m}{a}\right]_\alpha \\
& + \sum_{m=1}^{M}\sum_{\alpha=4}^{12}\left[\oint_{L_m}(\overline{\sigma}_{\alpha ij}U_{ik} - \overline{u}_{\alpha i}\Sigma_{ijk})\,\epsilon_{jnt}z_n dz_t\right] \times \\
& \quad \left[R_{,r}\left(\overset{m}{T}^{-1}\overset{*m}{a} + \left(\overset{m}{T}^{-1}\right)^*\overset{m}{a}\right)\right]_{\alpha-3} \\
& - \sum_{m=1}^{M}\sum_{\alpha=4}^{12}\left[\oint_{L_m}(\overline{\sigma}_{\alpha ij}U_{ik} - \overline{u}_{\alpha i}\Sigma_{ijk})\,\epsilon_{jrt}dz_t\right] \times
\end{aligned}
$$

$$\left[R \left(\overset{m}{T^{-1}} \overset{*m}{a} + \left(\overset{m}{T^{-1}} \right)^* \overset{m}{a} \right) \right]_{\alpha-3}$$

$$+ \sum_{m=1}^{M} \sum_{\alpha=1}^{3} \left[\oint_{L_m} D_{\alpha jk} dz_j \right] \left[S_{,r} \left(\overset{m}{T^{-1}} \overset{*m}{a} + \left(\overset{m}{T^{-1}} \right)^* \overset{m}{a} \right) \right]_{\alpha}$$

$$+ \sum_{m=1}^{M} \sum_{\alpha=1}^{3} \left[\oint_{L_m} \Sigma_{\alpha jk} \epsilon_{jrt} dz_t \right] \left[S \left(\overset{m}{T^{-1}} \overset{*m}{a} + \left(\overset{m}{T^{-1}} \right)^* \overset{m}{a} \right) \right]_{\alpha}$$

$$(5.63)$$

The curvatures $u_{k,rp}$ in the above equation can be obtained from (see [103]):

$$u_{k,rp}(\boldsymbol{\xi}) = \sum_{m=1}^{M} \sum_{\alpha=13}^{27} \left[\oint_{L_m} (\overline{\sigma}_{\alpha ij} U_{ik,r} - \overline{u}_{\alpha i} \Sigma_{ijk,r}) \epsilon_{jpt} dz_t \right] \left[\overset{m}{T^{-1}} \overset{m}{a} \right]_{\alpha}$$

$$- \sum_{m=1}^{M} \sum_{\alpha=4}^{12} \left[\oint_{L_m} (\overline{\sigma}_{\alpha ij} U_{ik} - \overline{u}_{\alpha i} \Sigma_{ijk}) \epsilon_{jrt} dz_t \right] \left[R_{,p} \overset{m}{T^{-1}} \overset{m}{a} \right]_{\alpha-3}$$

$$+ \sum_{m=1}^{M} \sum_{\alpha=4}^{12} \left[\oint_{L_m} (\overline{\sigma}_{\alpha ij} U_{ik,r} - \overline{u}_{\alpha i} \Sigma_{ijk,r}) \epsilon_{jpt} dz_t \right] \left[R \overset{m}{T^{-1}} \overset{m}{a} \right]_{\alpha-3}$$

$$+ \sum_{m=1}^{M} \sum_{\alpha=1}^{3} \left[\oint_{L_m} \Sigma_{\alpha jk} \epsilon_{jrt} dz_t \right] \left[S_{,p} \overset{m}{T^{-1}} \overset{m}{a} \right]_{\alpha}$$

$$- \sum_{m=1}^{M} \sum_{\alpha=1}^{3} \left[\oint_{L_m} \Sigma_{\alpha jk,r} \epsilon_{jpt} dz_t \right] \left[S \overset{m}{T^{-1}} \overset{m}{a} \right]_{\alpha}$$

$$+ \sum_{m=1}^{M} \sum_{\alpha=1}^{3} \left[\oint_{L_m} D_{\alpha jk} dz_j \right] \left[S_{,rp} \overset{m}{T^{-1}} \overset{m}{a} \right]_{\alpha} \qquad (5.64)$$

Finally, stress sensitivities can be easily obtained from $\overset{*}{v}_{kr}$ by using Hooke's law.

5.4   Numerical Results: Hollow Sphere

Sensitivity results are presented in this section for a thick hollow sphere under internal pressure (see Section 4.4). As before, the inner and outer radii of the sphere are 1 and 2 units, respectively, the shear modulus $G = 1$, the Poisson's ratio $\nu = 0.3$ and the internal pressure is 1 unit. The design variable is the inner radius a of the sphere.

A generic surface mesh on a one-eighth sphere is given in Figure 4.2. Three levels of discretization, coarse, medium and fine, are used in this work. Mesh statistics are shown in Table 4.3.

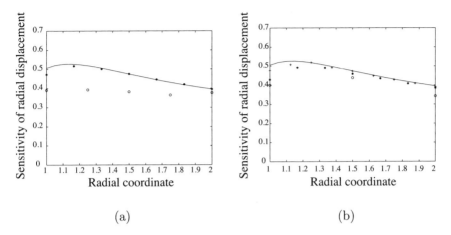

(a)   (b)

Figure 5.1: Sensitivity of radial displacement along (a) the x_3 axis and (b) along the line $x_1 = 0$, $x_2 = x_3$. Exact solution —. Numerical solutions: coarse mesh ○ ○ ○○, medium mesh ∗ ∗ ∗∗, fine mesh ++++ (from [100])

5.4.1 Sensitivities on Sphere Surface

5.4.1.1 Displacement sensitivities

Displacement sensitivities (from equation (5.34)), along various lines on the sphere surface, for different discretizations, appear in Figure 5.1. (A typical displacement profile for this example appears in Figure 4.3.) A linear design velocity profile:

$$\overset{*}{R} = \frac{b - R}{b - a} \qquad (5.65)$$

is used here. It is seen that the numerical results for the coarse mesh exhibit large errors. The reason for this is under investigation. However, they do appear to converge to the exact solution with increasing mesh density. Please see Chandra and Mukherjee [22] for a discussion of analytical solutions for design sensitivities for various examples.

5.4.1.2 Stress sensitivities

Stress sensitivities on the surface of the sphere are calculated from equations (5.38 - 5.39). (Stress profiles for this example appear in Figure 4.4). Numerical solutions from the medium mesh, for stress sensitivities on the outer surface $R = b$, agree well with the exact solution (Figure 5.2).

The situation, however, is somewhat tricky on the inner surface $R = a$. Here, one has:

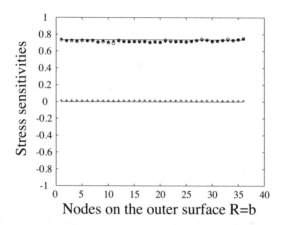

Figure 5.2: Stress sensitivities on the outer surface $R = b$. Exact solutions —.
Numerical solutions from the medium mesh: $\overset{*}{\sigma}_{RR}$ + + + +, $\overset{*}{\sigma}_{\theta\theta}$ ○○○○, $\overset{*}{\sigma}_{\phi\phi}$ ****
(from [100])

$$\overset{*}{\sigma}_{\theta\theta} = \overset{\triangle}{\sigma}_{\theta\theta} + \frac{\partial \sigma_{\theta\theta}}{\partial R} \overset{*}{R} \qquad (5.66)$$

(and similarly for the other components of stress). The exact solution for $\overset{\triangle}{\sigma}_{\theta\theta} = \overset{\triangle}{\sigma}_{\phi\phi}$ is $120/49$, which is positive, while the convected term in equation (5.66) is $-12/7$, which is negative. As can be seen from Figures (5.3-5.4), this situation makes it quite difficult to calculate $\overset{*}{\sigma}_{\theta\theta}$ and $\overset{*}{\sigma}_{\phi\phi}$ accurately, especially when accuracy is measured in terms of percentage errors, because one must calculate the difference between two numbers that are reasonably close.

Table 5.1 shows percentage root mean square errors in $\overset{*}{\sigma}_{\theta\theta}$ and $\overset{*}{\sigma}_{\phi\phi}$, respectively. These are defined as:

$$\epsilon = \frac{100}{f_{exact}} \sqrt{\frac{1}{n} \sum_{i=1}^{n} (f_{exact} - f_{i\ numer})^2} \qquad (5.67)$$

It is seen that while the errors on the outer surface are very low, those on the inner surface, even with the fine mesh, are quite high. Perhaps further work on this problem, including the development of a different algorithm for calculation of surface stress sensitivities, needs to be carried out.

5.4.2   Sensitivities at Internal Points

5.4.2.1   Sensitivity of radial displacement

Figure 5.5(a) shows the sensitivity of radial displacement along the line $x_1 = x_2 = x_3$. These results are obtained from equation (5.54) which is used after first

		medium mesh	fine mesh
$R = b$	$\overset{*}{\sigma}_{\theta\theta}$	1.99	1.62
	$\overset{*}{\sigma}_{\phi\phi}$	1.77	2.01
$R = a$	$\overset{*}{\sigma}_{\theta\theta}$	25.45	15.15
	$\overset{*}{\sigma}_{\phi\phi}$	26.49	14.88

Table 5.1: Percentage root mean square errors in sensitivities of stress components (from [100])

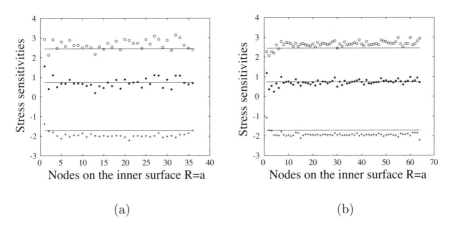

(a)   (b)

Figure 5.3: Stress sensitivities on the inner surface $R = a$. Exact solutions: —. Numerical solutions from (a) the medium mesh and (b) the fine mesh: $\overset{\triangle}{\sigma}_{\theta\theta}$ ∘ ∘ ∘∘, $\frac{\partial \sigma_{\theta\theta}}{\partial R} \overset{*}{R}$ + + + +, $\overset{*}{\sigma}_{\theta\theta}$ * * ** (from [100])

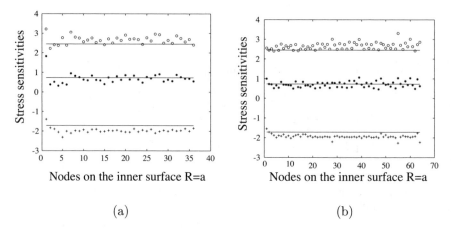

(a)   (b)

Figure 5.4: Stress sensitivities on the inner surface $R = a$. Exact solutions: —. Numerical solutions from (a) the medium mesh and (b) the fine mesh: $\overset{\triangle}{\sigma}_{\phi\phi}$ ○ ○ ○○, $\frac{\partial \sigma_{\phi\phi}}{\partial R} \overset{*}{R}$ + + + +, $\overset{*}{\sigma}_{\phi\phi}$ * * ** (from [100])

solving the sensitivity boundary value problem (5.34). Again, the linear velocity profile (5.65) is used here. It is seen from Figure 5.5(a) that the agreement between the numerical and analytical solutions is very good.

5.4.2.2   Sensitivities of stresses

Sensitivities of the stress components $\sigma_{\theta\theta}$ and σ_{RR}, along the line $x_1 = x_2 = x_3$, are presented in Figure 5.5(b). Again, very good agreement is observed between the numerical and analytical solutions.

5.5   Numerical Results: Block with a Hole

5.5.1   Geometry and Mesh

This example is concerned with a rectangular block with a cylindrical hole of circular cross-section, loaded in uniform remote tension. The BCM model is fully three-dimensional but the imposed boundary conditions are chosen such that a state of plane strain prevails in the block and the numerical results obtained from the BCM can be compared with Kirsch's analytical solution for the corresponding 2-D plane strain problem. Of course, the analytical solution is only available for an infinitesimal hole in a slab and this is referred to as the "exact" solution in this section of this chapter.

The side of a square face of the full block is 20 units, the hole diameter is 2 units and the thickness (in the x_3 direction) is 6 units. One-eighth of the block is modeled in order to take advantage of symmetry. The mesh on the front and

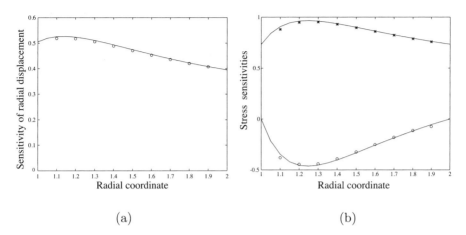

(a)   (b)

Figure 5.5: (a) Sensitivity of radial displacement along the line $x_1 = x_2 = x_3$. Exact solution —. Numerical solution from the fine mesh: ○○○○. (b) Sensitivity of stresses along the line $x_1 = x_2 = x_3$. Exact solutions —. Numerical solutions from the fine mesh: $\overset{*}{\sigma}_{\theta\theta}$ ★★★★ $\overset{*}{\sigma}_{RR}$ ○○○○ (from [103])

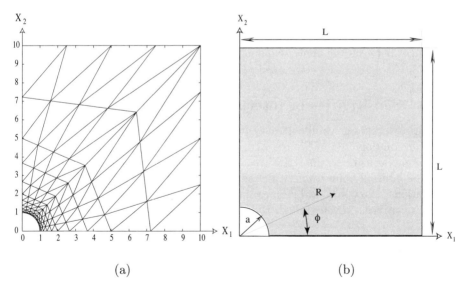

(a)   (b)

Figure 5.6: (a) Mesh on a quarter of the front and back faces of a rectangular block with a circular cylindrical hole. (b) Design velocities for this example (from [103])

back faces ($x_3 = 3$ and $x_3 = 0$, respectively) of the one-eighth block is shown in Figure 5.6(a). Four layers of triangles (in the thickness direction) constitute the mesh on the remaining surfaces of the one-eighth block. The complete block is loaded by uniform remote tensions in the x_1 and x_2 directions while $u_3 = 0$ on the faces normal to the x_3 axis in order to simulate plane strain conditions. Of course, boundary conditions on the symmetry planes are applied in the computer model of the one-eighth block in the usual way.

The design variable in this example is the hole radius a. The chosen design velocity distribution in the slab is linear along any radial direction in any square section normal to the x_3 axis and is independent of the x_3 coordinate. In other words, referring to Figure 5.6(b), one has:

$$\overset{*}{R} = \frac{L/\cos\phi - R}{L/\cos\phi - a} \quad \text{for} \quad \phi < \pi/4 \tag{5.68}$$

and

$$\overset{*}{R} = \frac{L/\sin\phi - R}{L/\sin\phi - a} \quad \text{for} \quad \phi \geq \pi/4 \tag{5.69}$$

5.5.2 Internal Stresses

Comparisons of numerical and analytical solutions for internal stresses in the block are presented in Figures 5.7(a) and (b). Two cases are considered: stresses along the line $x_1 = x_2$, $x_3 = 1.5$ for uniaxial remote loading in the x_1 direction, and along the line $x_1 = \sqrt{3}x_2$, $x_3 = 1.5$ for equal biaxial loading. The agreement between the analytical and numerical results is again seen to be very good.

5.5.3 Sensitivities of Internal Stresses

This last example is a difficult one. It is concerned with stress sensitivities for the situation depicted in Figure 5.7(a).

The numerical and analytical results for this problem are shown in Figure 5.8. The internal point numerical results, which are obtained from independent calculations, show acceptable agreement with the analytical solutions for the stress sensitivities, even though the analytical solutions vary rapidly near the hole.

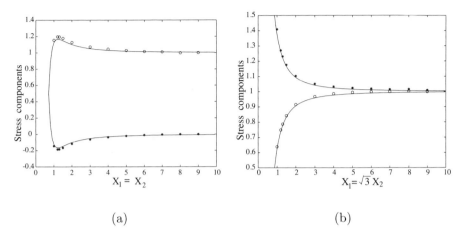

(a)   (b)

Figure 5.7: Stresses (a) along the line $x_1 = x_2$, $x_3 = 1.5$ for uniaxial loading in the x_1 direction and (b) along the line $x_1 = \sqrt{3}x_2$, $x_3 = 1.5$ for equal biaxial loading. Exact solutions —. Numerical solutions: σ_{11} ∘∘∘∘, σ_{22} ∗∗∗∗ (from [103])

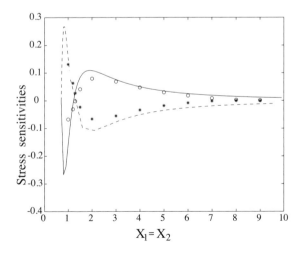

Figure 5.8: Stress sensitivities along the line $x_1 = x_2$, $x_3 = 1.5$ for uniaxial loading in the x_1 direction. Exact solutions: $\overset{*}{\sigma}_{11}$ — , $\overset{*}{\sigma}_{22}$ − − −. Numerical solutions: $\overset{*}{\sigma}_{11}$ ∘∘∘∘, $\overset{*}{\sigma}_{22}$ ∗∗∗∗ (from [103])

Chapter 6

SHAPE OPTIMIZATION

Shape optimization of 3-D elasticity problems, with the BCM, is the subject of this chapter. Further details are available in Shi and Mukherjee [150].

6.1 Shape Optimization Problems

An optimal shape design problem can be stated as a minimization problem under certain constraints. Its general form can be stated as follows:

$$\text{Minimize} \quad f(\mathbf{b}) \tag{6.1}$$

$$\text{Subject to} \quad g_i(\mathbf{b}) \geq 0, \qquad i = 1, ..., N_g \tag{6.2}$$

$$h_j(\mathbf{b}) = 0, \qquad j = 1, ..., N_h \tag{6.3}$$

$$b_k^{(\ell)} \leq b_k \leq b_k^{(u)}, \quad k = 1, ..., N \tag{6.4}$$

in which $\mathbf{b} = \langle b_1, b_2, ..., b_N \rangle^T$ are the design variables, $f(\mathbf{b})$ is the objective function, and $g_i(\mathbf{b})$ and $h_j(\mathbf{b})$ are the inequality and equality constraints, respectively. Equation (6.4) gives side constraints that are used to limit the search region of an optimization problem. Here, the parameters $b_k^{(\ell)}$ and $b_k^{(u)}$ denote the lower and upper bounds, respectively, of the design variable b_k.

The most common mathematical programming approaches, used in gradient-based optimization algorithms, are the successive linear programming (SLP) and successive quadratic programming (SQP) methods. In the SQP method, the optimization problem is approximated by expanding the objective function in a second order Taylor series about the current values of the design variables, while the constraints are expanded in a first order Taylor series.

The subroutine DNCONF from the IMSL library is coupled with a 3-D BCM code for elastic stress analysis in order to carry out the shape optimization examples that are described in this chapter. This subroutine, that uses the SQP method, is based on the FORTRAN subroutine NLPQL developed by Schittkowski [146]. The required gradients of the objective functions, constraints etc. are calculated internally by the optimization subroutine by the finite difference method. Of course, these sensitivities could also have been obtained by the direct differentiation approach described in Chapter 5.

6.2   Numerical Results

Two illustrative shape optimization examples, in 3-D linear elasticity, are solved using the BCM coupled with the IMSL optimization subroutine mentioned above. The first example is that of optimizing the shape of a fillet in a tension bar whose volume is selected as the optimization function. An optimal shape is sought that minimizes the volume (and therefore the weight when the bar material has constant density) without causing yielding anywhere in the bar. This is the axisymmetric version of the planar problem described in Phan et al. [132] - Section 5.3. Of course, the full 3-D BCM code is used here.

The second problem is concerned with a cube with an ellipsoidal cavity loaded in remote triaxial tension. An interesting result in 2-D [5, 132], for an infinite elastic plate with an elliptical cutout, loaded in remote biaxial tension, is as follows. Let the semimajor and semiminor axes of the elliptical hole be a_1 and a_2, respectively, and the coordinate axes be centered at the center of the ellipse with the x_1 and x_2 directions along the major and minor axes, respectively. Also, let S_1 and S_2 be the remote tensile loadings in the coordinate directions. Now, if $a_2/a_1 = S_2/S_1$, then the tangential stress around the elliptical cutout is uniform ! This problem has been solved in 2-D by the BCM in [132] and an elastic-plastic version of this 2-D problem has been solved in [169] (see, also, [22]) with the BEM. The 3-D elasticity problem is described in this chapter and some interesting results are obtained.

6.2.1   Shape Optimization of a Fillet

The initial cross-section of the axisymmetric bar is shown in Figure 6.1. The Young's modulus, Poisson's ratio and allowable von Mises stress are taken as $E = 10^7 psi$, $\nu = 0.3$ and $\hat{\sigma}^{(VM)} = 120$ psi, respectively. The bar is loaded by a uniform axial tensile traction of 100 psi . The design surface is the surface of revolution (initially a cone) obtained by revolving the curve ED about the symmetry axis AB. The end circles of this surface of revolution are kept fixed. The variable boundary ED is modeled as a cubic spline (using the IMSL subroutine DCSDEC) which is defined by the fixed end points E and D and by the variable points $C_k, k = 1, 2, 3$. The points E, $C_k, k = 1, 2, 3$ and D have equally spaced projections on the axis of symmetry while the radii $r_k, k = 1, 2, 3$ are the design variables.

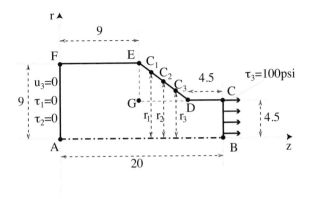

Figure 6.1: Modeling of a bar with a fillet (from [150])

Quadratic CIM9 elements (see Figure 4.1) are used in both the numerical examples described in this chapter. A total of 160 elements are used to discretize the surface of the bar. These are distributed as follows. There are 16 elements on each of the end circles (on the planes $z = 0$ and $z = 20$, respectively) of the bar. The surface of revolution described by the edge FE has 32 elements, the one described by DC has 32 elements, and the design surface has 64 elements, respectively.

The objective function is the volume of the axisymmetric object bounded by the plane $z = 9$, the cylindrical surface $r = 4.5$ and the design surface. This is:

$$\phi(r_1, r_2, r_3) = \int_{z_G}^{z_D} \pi \left[r^2(r_1, r_2, r_3, z) - (4.5)^2 \right] dz \qquad (6.5)$$

The axisymmetric shape of the body is maintained during the optimization process and the design nodes C_k, $k = 1, 2, 3$ are constrained to lie within the triangle EGD. In addition, one has:

$$\sigma_i^{(VM)}/\hat{\sigma}^{(VM)} \leq 1.0, \qquad 1 \leq i \leq n_s \qquad (6.6)$$

where $\sigma_i^{(VM)}$ are the values of the von Mises stress at the centroids of certain elements on the curved surface of the bar. Here, since the physical problem is axisymmetric, the elements chosen are the ones whose centroids lie along FEDC, so that $n_s = 16$.

The usual definition of the von Mises stress is used. This is:

$$[\sigma^{(VM)}]^2 = (3/2)s_{ij}s_{ij} \qquad (6.7)$$

in terms of s_{ij}, the deviatoric components of the stress σ_{ij}.

The final converged solution, which is obtained after just 1 iteration, is shown in Figure 6.2(a) while Figure 6.2(b) shows the objective function as a

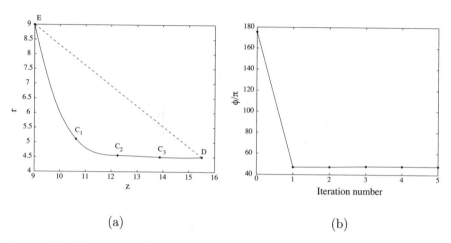

(a)   (b)

Figure 6.2: (a) Optimal shape of the fillet (b) Objective function ϕ as a function of iteration number (from [150])

function of iteration number. The initial value of the objective function is 551.35 with one stress constraint being violated. The final (converged) value is 150.39 when two stress constraints are active. Thus, the final volume of the design portion of the bar is about 27.28 % of its initial value. In the corresponding 2-D problem [132], the design area reduces to 49.85% of its initial value.

6.2.2   Optimal Shapes of Ellipsoidal Cavities Inside Cubes

A cube with a centrally located ellipsoidal cavity, loaded in remote triaxial tension, is shown in Figure 6.3. Of course, the faces of the cube, which are not shown in this figure, are suitably restrained. Consistent units are used. This time the shear modulus is taken as $G = 10^5$ and the Poisson's ratio $\nu = 0.3$. The cube is of size $30 \times 30 \times 30$. The cavity surface is the design surface with the ellipsoid semi axes a_1 and a_2 as the design variables ($a_3 = 1$). The cube surface is fixed. Three cases of remote loadings (cases (1), (2) and (3)) are considered : $S_1 : S_2 : S_3 = 1 : 1 : 1$; $2 : 2 : 1$ and $2 : 1.5 : 1$, respectively, with $S_3 = 10^5$ in all cases.

The mesh consists of 8 identical CIM9 elements on each surface of the cube and 72 CIM9 elements on the surface of the cavity. The cavity mesh evolves with the changing shape of the cavity during the optimization process.

As stated at the start of this section, the objective here is to find, if possible, the shape of an ellipsoidal cavity that would have uniform stress (i.e. some suitable measure of stress) on its surface for a given remote loading. As a test case, the first case of remote loading, i.e. $S_1 : S_2 : S_3 = 1 : 1 : 1$, is applied. Starting with an ellipsoidal cavity with axis ratios $a_1 : a_2 : a_3 = 3 : 2 : 1$, the cavity shape is sought that would make *all the stress components* on it uniform. This, of course, is a spherically symmetric problem, and, as expected,

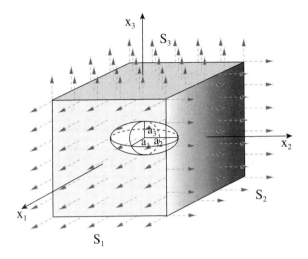

Figure 6.3: A cube with an ellipsoidal cavity under remote loading (from [150])

the optimal shape of the cavity is found to be a sphere.

The more interesting cases, of course, are loading cases (2) and (3). This time, two different scalar measures of stress on the cavity surface are considered. The first is σ_{kk}, the trace of the stress tensor, and the second is the von Mises stress. Uniform σ_{kk} on a surface means that $\sigma_{nn} + \sigma_{ss} + \sigma_{tt}$ (where $\mathbf{n}$ is normal and s and t are any two orthogonal directions, tangential to the surface at a point on it) is also uniform on the surface. Since the cavity surface is unloaded, this implies that $\sigma_{ss} + \sigma_{tt}$ is uniform for all points on the cavity surface. The criterion σ_{kk} *is uniform on a cavity surface* is treated as a generalization of the 2-D case in which the tangential stress on a cutout surface was made uniform [132].

The corresponding objective functions to be minimized are defined as:

$$\phi_1(a_1, a_2) = \frac{1}{n_s} \sum_{i=1}^{n_s} \left[\sigma_{kk}^{(i)} - \overline{\sigma_{kk}} \right]^2 \tag{6.8}$$

$$\phi_2(a_1, a_2) = \frac{1}{n_s} \sum_{i=1}^{n_s} \left[\left(\sigma_i^{(VM)} \right)^2 - \overline{\left(\sigma^{(VM)} \right)^2} \right]^2 \tag{6.9}$$

In the above equations, n_s is the number of elements on the design surface, a superscript (i) denotes evaluation of the appropriate quantity at the centroid of the i_{th} element on the design surface, and an overbar denotes the mean value of the appropriate quantity over the design surface.

Load case	Loads	Ellipse semiaxes	Objective function
Uniform σ_{kk} on cavity surface			
(2)	2, 2, 1	Start 2, 2, 1 u.b. 3,3,1	1.044
		Optimal 2.654, 2.654, 1	0.0027
(3)	2, 1.5, 1	Start 2, 1.5, 1 u.b. 7,7,1	0.11057
		Optimal 3.8022, 1.5608, 1	0.002286
Uniform $\sigma^{(VM)}$ on cavity surface			
(2)	2, 2, 1	Start 2, 2, 1 u.b. 9,9,1	3.4086
		Optimal 3.4105, 3.4105, 1	0.0489

Table 6.1: Results for cube with ellipsoidal cavity under remote loading. Upper bound is denoted as u.b. Lower bounds for ellipse semiaxes are 1,1,1. The loads are normalized with $S_3 = 1$ (from [150])

6.2.2.1   Uniform trace of the stress tensor over the cavity surface

The load cases (2) and (3) are tried for the first objective function ϕ_1. In each case, inspired by the 2-D case, the starting geometry of the ellipsoid is taken to be compatible with the loads. The results appear in Table 6.1 in which the side constraints on the design variables, as well as the initial and optimal values of the design variables and the objective function, are shown.

Details are shown in Figures 6.4 - 6.5. Figure 6.4(a) shows the stress σ_{kk}, and its mean value, initially and at the end of the optimization process. It is seen that the initial distribution of σ_{kk} (dots) is quite widespread while the optimal distribution (asterisks) is concentrated within a band around its mean value $\overline{\sigma_{kk}}$. The corresponding objective function, as a function of iteration number, is shown in Figure 6.4(b). It can be seen from Table 6.1 that, while the optimal values of a_1 and a_2 are equal (i.e. the cross-section of the optimal shaped cavity in the $x_3 = 0$ plane is a circle), the loading and geometry are not compatible as in the 2-D case (i.e. the optimal $a_1 : a_2 : a_3$ does not match the load ratios).

The more difficult case is load case (3). The stress σ_{kk} and the objective function for this case appear in Figures 6.5(a) and 6.5(b), respectively. Again, it can be seen from Figure 6.5(a) that the initial distribution of σ_{kk} (dots) is quite widespread while the optimal distribution (asterisks) is concentrated within a band around its mean value $\overline{\sigma_{kk}}$. It is seen from Figure 6.5(b) that in this case the behavior of the objective function is initially oscillatory. It does, however, eventually settle down to about 2.07% of its initial value. Examination of Table 6.1 again shows that the optimal cavity geometry is not compatible with the loading in this case.

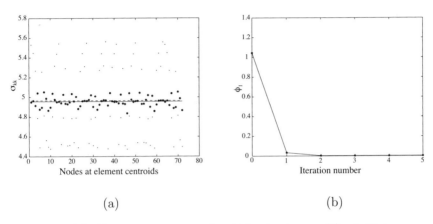

(a)   (b)

Figure 6.4: (a) The stresses σ_{kk} and $\overline{\sigma_{kk}}$ for load case (2). Initial : σ_{kk} ...,
$\overline{\sigma_{kk}}$ − − − Optimal : σ_{kk} * **, $\overline{\sigma_{kk}}$ — (b) Objective function ϕ_1 as a function of
iteration number for load case (2) (from [150])

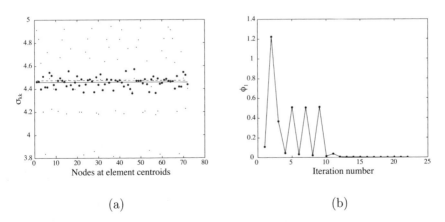

(a)   (b)

Figure 6.5: (a) The stresses σ_{kk} and $\overline{\sigma_{kk}}$ for load case (3). Initial : σ_{kk} ...,
$\overline{\sigma_{kk}}$ − − − Optimal : σ_{kk} * **, $\overline{\sigma_{kk}}$ — (b) Objective function ϕ_1 as a function of
iteration number for load case (3) (from [150])

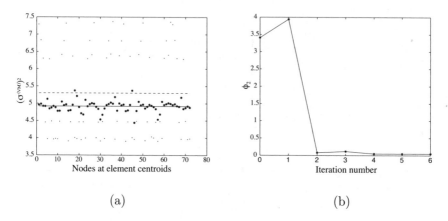

(a)   (b)

Figure 6.6: (a) The stresses $(\sigma^{(VM)})^2$ and $\overline{(\sigma^{(VM)})^2}$ for load case (2). Initial : $(\sigma^{(VM)})^2$ ..., $\overline{(\sigma^{(VM)})^2}$ - - - Optimal : $(\sigma^{(VM)})^2$ * **, $\overline{(\sigma^{(VM)})^2}$ — (b) Objective function ϕ_2 as a function of iteration number for load case (2) (from [150])

6.2.2.2   Uniform von Mises stress over the cavity surface

The final example considers the load case (2) with the second objective function ϕ_2. The results are summarized in Table 6.1 and details are shown in Figures 6.6(a) and 6.6(b). This time the converged value of the objective function is 1.43% of its initial value. Figure 6.6(a) shows a significant difference between the initial and final values of the mean value of $(\sigma^{(VM)})^2$. Also, the optimal distribution of $(\sigma^{(VM)})^2$ (asterisks) lies in a wider band around its mean value compared to the situations in Figures 6.4(a) and 6.5(a) . As expected, the optimal cavity shape has a circle in the $x_3 = 0$ plane.

The case of ϕ_2 with the load case (3) (2:1.5:1) did not converge. It is conjectured that a more general cavity shape, rather than an ellipsoidal one, needs to be considered for this difficult problem.

6.2.3   Remarks

The optimal shape of a fillet in an axially loaded bar is obtained quickly within just one iteration ! The size of the design region reduces to nearly a quarter of its initial value in this 3-D problem, compared to about half in the corresponding 2-D problem.

The next example considered is the optimal shape of an ellipsoidal cavity in a cube loaded by remote tensions. The sought after cavity shape is the one that makes some measure of the stress uniform on it. It is known that in the corresponding 2-D problem, for an elliptical cutout in an infinite plate, the tangential stress on the cutout surface is uniform when the loading and geometry are compatible. (The precise meaning of compatibility is explained in the body of this chapter). It is found that such is not the case in the 3-D problem. Also, optimization of this class of problems is often more difficult than for the previous bar with a fillet example. This is true if either no two of the remote tractions are equal, and/or if one tries to make the von Mises stress uniform on the cutout surface. Sometimes (Figure 6.5(b)) one sees initial oscillations of the objective function as a function of iteration number.

Chapter 7

ERROR ESTIMATION AND ADAPTIVITY

The subject of this chapter is error analysis and adaptivity with the BCM. The idea of using hypersingular residuals, to obtain local error estimates for the BIE, was first proposed by Paulino [122] and Paulino et al. [123]. This idea has been applied to the collocation BEM (Paulino et al. [123], Menon et al. [96] and Paulino et al. [127]); and has been discussed in detail in Chapter 2 of this book. The main idea, applied to the BCM, has appeared in Mukherjee and Mukherjee [111], and is presented in this chapter.

7.1 Hypersingular Residuals as Local Error Estimators

The usual BCM equation (4.25) is solved first for the boundary variables (tractions and displacements) $\mathbf{a}$. Next, this value of $\mathbf{a}$ is input into the right-hand side of equation (4.44) in order to obtain the hypersingular residuals v_{kn} in the displacement gradients $u_{k,n}$. Next, the stress residuals are obtained from Hooke's law:

$$s_{kn} = \lambda v_{mm}\delta_{kn} + \mu(v_{kn} + v_{nk}) \tag{7.1}$$

Finally, a scalar measure r of the residual, evaluated at the centroid of a triangular surface element, is postulated based on the idea of energy. This is:

$$r = s_{kn}v_{kn} \tag{7.2}$$

It has been proved in [96] and [127] for the BIE that, under certain favorable conditions, real positive constants c_1 and c_2 exist such that:

$$c_1 r \leq \epsilon \leq c_2 r \tag{7.3}$$

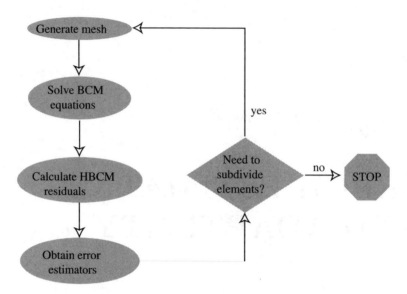

Figure 7.1: Flow chart for adaptive meshing (from [111])

where r is some scalar measure of a hypersingular residual and ϵ is a scalar measure of the exact local error. Thus, a hypersingular residual is expected to provide a good estimate of the local error on a boundary element. It should be mentioned here that the definitions of the residuals used in [96] and [127] are analogous to, but different in detail from, the ones proposed in this chapter.

In the rest of this chapter, $e = r$, where r, defined in equation (7.2) (and evaluated at an element centroid), is the hypersingular residual, and e is the local element error estimator that is used to drive an h-adaptive procedure with the BCM.

7.2   Adaptive Meshing Strategy

The flow chart for adaptive meshing is shown in Figure 7.1.

The remeshing strategy is based on the values of the error estimator e at each element centroid. This strategy is shown in Figure 7.2 in which $\bar{e}$ is the average value of the error estimator e over all the boundary elements.

A possible criterion for stopping cell refinement can be:

$$\bar{e} \leq e_{global} \qquad (7.4)$$

where e_{global} has a preset value that depends on the level of overall desired accuracy.

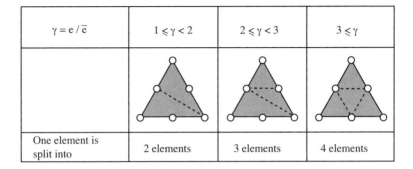

$\gamma = e/\overline{e}$	$1 \leqslant \gamma < 2$	$2 \leqslant \gamma < 3$	$3 \leqslant \gamma$
One element is split into	2 elements	3 elements	4 elements

Figure 7.2: Remeshing strategy (from [111])

7.3 Numerical Results

7.3.1 Example One - Short Clamped Cylinder under Tension

This first example is concerned with a short cylinder which is clamped at the bottom and subjected to unit tensile traction on the top surface (Figure 7.3(a)). The radius and length of the cylinder are each 2 units, the shear modulus of the cylinder material is 1.0 and the Poisson's ratio is 0.3 (in consistent units). The initial mesh on the top (loaded) and bottom (clamped) faces of the cylinder are identical and are shown in Figure 7.3 (b) while the initial uniform mesh on its curved surface is shown in Figure 7.4 (b).

It is known ([37], [137]) that, for this problem, the normal stress component σ_{33} varies slowly over much of the clamped face, but exhibits sharp gradients near its boundary. This stress component becomes singular on the boundary of the clamped face. The behavior of the shearing stress component σ_{zr} (here $r, \theta, z \equiv 3$ are the usual polar coordinates) on the clamped face is qualitatively similar to that of σ_{33}. The stresses are uniform on the loaded face.

It is seen from Figures 7.3 and 7.4 that this behavior is captured well by the adaptive scheme. Element error estimators are obtained from equations (4.44), (7.1) and (7.2) after first averaging the traction results from (4.25) within each element and then using these averaged traction values. Figure 7.3 (b) shows that these element error estimators (denoted by vertical bars at element centroids) are largest on the elements near the boundary of the clamped face. As a consequence (Figures 7.3 (b), (c), (d)) the region near the boundary of the clamped face is refined most while the mesh on the loaded face of the cylinder is left unaltered. Also, Figures 7.4 (c), (d) show that some mesh refinement

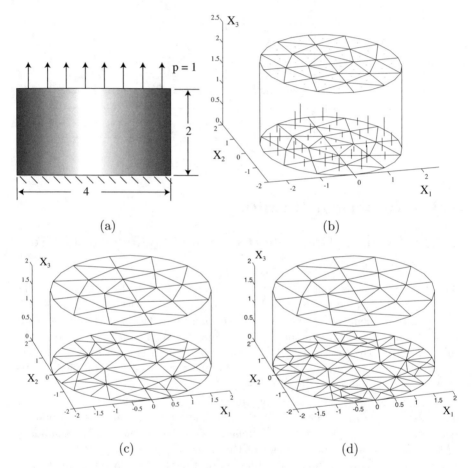

Figure 7.3: Adaptive meshing of the top and bottom faces of a clamped cylinder under tension: (a) geometry and loading (b) initial mesh with element error estimators (c) mesh at the end of the first adaptive step (d) mesh at the end of the second adaptive step (from [111])

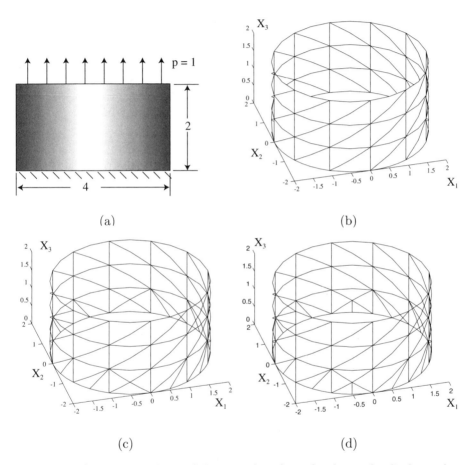

Figure 7.4: Adaptive meshing of the curved surface of a clamped cylinder under tension: (a) geometry and loading (b) initial uniform mesh (c) mesh at the end of the first adaptive step (d) mesh at the end of the second adaptive step (from [111])

mesh	# of elements	# of nodes	$\bar{e}$
initial	144	290	0.0086799
after 1st adaptive step	192	386	0.0048994
after 2nd adaptive step	246	494	0.0042723

Table 7.1: Mesh statistics and $\bar{e}$ for the clamped cylinder under tension (from [111])

takes place on the bottom layer of the curved surface of the cylinder, which is nearest to the clamped face, while the rest of the mesh on it remains unaltered.

Finally, the mesh statistics, together with $\bar{e}$, the average value of the error estimator e over the entire surface of the cylinder, appear in Table 7.1. As expected, $\bar{e}$ is seen to decrease with mesh refinement.

7.3.2   Example Two - the Lamé Problem for a Hollow Cylinder

This example is concerned with a thick hollow cylinder, in plane strain, subjected to external radial tensile loading. The inner and outer radii of the hollow cylinder are 1 and 3 units, respectively. The shear modulus of the material is 1.0, the Poisson's ratio is 0.3 and the external radial traction is 3 (in consistent units). A quarter of the cylinder is modeled and the initial mesh on the quarter cylinder is shown in Figure 7.5. The bars in Figure 7.5 are the error estimators evaluated at the centroids of the boundary elements. As expected ([124], also, please see the discussion in the following paragraph), the error estimators are largest on the surface of the hole and on the elements on the upper and lower surfaces (EBAF and HCDG) of the cylinder that lie near the hole. (The visible elements are shown in Figure 7.5 and the hidden ones are not).

The next (and final) mesh, obtained from the adaptive strategy outlined in Section 7.2 above, is shown in Figure 7.6. It is seen that mesh refinement is carried out vigorously on the upper and lower surfaces EBAF and HCDG of the cylinder (the hidden elements are not shown in Figure 7.6), as well as on the surface FADG of the hole, while the symmetry planes ABCD and EFGH, on which the stresses are independent of the x_3 coordinate, are only slightly refined in order to maintain mesh compatibility. Of course, refinement of the surfaces EBAF and HCDG is expected in view of the presence of radial stress gradients on these surfaces. The situation on the curved surface FADG is particularly interesting. In this axisymmetric problem, the tangential gradients of the stress fields in the θ direction are, of course, always zero. It is important to note, however, that the radial stress gradients are large at points on the hole surface, and this fact leads to large error estimators and significant refinement of the boundary elements on the surface FADG. The corresponding 2-D case is discussed, in some detail, in Paulino et al. [124].

It is important to check the behavior of the actual errors, when the exact

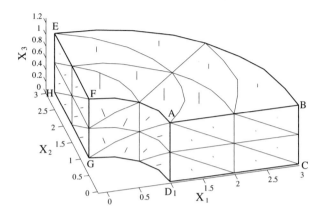

Figure 7.5: Lamé problem - initial mesh on quarter cylinder together with element error estimators (from [111])

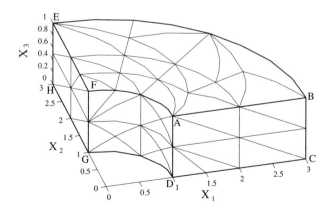

Figure 7.6: Lamé problem - final mesh on quarter cylinder (from [111])

solution is available, in adaptive meshing problems such as this example. The tangential stress $\sigma_{\theta\theta}$, as a function of the radial distance r from the center of the cylinder, is shown in Figure 7.7. The solid line in Figure 7.7 is the exact solution (from, e.g. [167]) while the numerical results, from the initial and the final mesh, are designated by open circles and plus signs, respectively. The numerical results for the tangential stress are obtained from the calculated tractions at the traction nodes I_i (see Figure 4.1) on the boundary elements on the symmetry face ABCD in Figures 7.5 and 7.6. The inaccurate results from the initial coarse mesh is a consequence of the chosen mesh, not the method itself. This can be seen, for example, by observing the BCM results for the Lamé problem for a hollow sphere under internal pressure in Figure 4.3 (open circles), obtained from a reasonably fine mesh; as well as by examining other numerical results from the BCM, in, for example, Chapter 4 (see, also, Mukherjee et al. [109]).

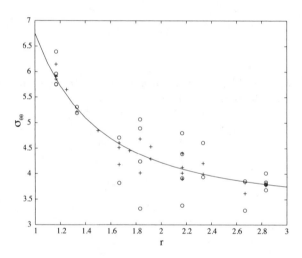

Figure 7.7: Lamé problem for a hollow cylinder. Tangential stress $\sigma_{\theta\theta}$ as a function of radial distance r. Exact solution: ——, BCM solution from initial mesh: o o o o, BCM solution from final mesh: ++++ (from [111])

The $\mathcal{L}_2$ error in a numerical solution in Figure 7.7 is defined as:

$$\epsilon = \frac{100}{\bar{\sigma}_{\theta\theta}} \sqrt{\frac{\sum_{i=1}^{n}(\epsilon_i)^2}{n}} \qquad (7.5)$$

where the pointwise error $\epsilon_i = (\sigma_{\theta\theta})^{(i)}_{numerical} - (\sigma_{\theta\theta})^{(i)}_{exact}$ at node i, n is the number of nodes and $\bar{\sigma}_{\theta\theta}$ is the average value of the exact solution for $\sigma_{\theta\theta}$ (here 4.5). The resulting values of the $\mathcal{L}_2$ errors are 9.83% and 3.83% for the initial and final mesh, respectively. The adaptive meshing procedure is seen to reduce the error significantly in one step.

Part III

THE BOUNDARY NODE METHOD

Chapter 8

SURFACE APPROXIMANTS

A moving least squares (MLS) approximation scheme, using curvilinear coordinates on the 1-D bounding surface of a 2-D body, or on the 2-D bounding surface of a 3-D body, is suitable for the BNM. The 2-D problem, which uses the curvilinear coordinate s on the boundary of a body, is discussed in detail in Mukherjee and Mukherjee [107] (potential theory) and in Kothnur et al. [72] (elasticity) (see, also, [108, 52, 77]). The 3-D problem requires the curvilinear surface coordinate $\mathbf{s}$ with components (s_1, s_2). (Chati and Mukherjee [26], potential theory; Chati et al. [25], elasticity). This procedure is described below. A brief discussion, of ongoing work on the BNM with Cartesian coordinates [79, 80, 163, 164], is presented as well.

8.1 Moving Least Squares (MLS) Approximants

It is assumed that, for 3-D problems, the bounding surface ∂B of a solid body is the union of piecewise smooth segments called panels. On each panel, one defines surface curvilinear coordinates (s_1, s_2). For problems in potential theory, let u be the unknown potential function and $\tau \equiv \partial u / \partial n$ (where $\mathbf{n}$ is a unit outward normal to ∂B at a point on it). For 3-D linear elasticity, let u denote a component of the displacement vector $\mathbf{u}$ and τ be a component of the traction vector $\boldsymbol{\tau}$ on ∂B. One defines :

$$u(\mathbf{s}) = \sum_{i=1}^{m} p_i(\mathbf{s} - \mathbf{s}^E)a_i = \mathbf{p}^T(\mathbf{s} - \mathbf{s}^E)\mathbf{a}$$

$$\tau(\mathbf{s}) = \sum_{i=1}^{m} p_i(\mathbf{s} - \mathbf{s}^E)b_i = \mathbf{p}^T(\mathbf{s} - \mathbf{s}^E)\mathbf{b} \tag{8.1}$$

The monomials p_i (see below) are evaluated in local coordinates $(s_1 - s_1^E, s_2 - s_2^E)$ where (s_1^E, s_2^E) are the global coordinates of an evaluation point E. It is important to state here that a_i and b_i are not constants. Their functional dependencies are determined later. (The name "moving least squares" arises from the fact that the quantities a_i and b_i are not constants). The integer m is the number of monomials in the basis used for u and τ. Quadratic interpolants, for example, are of the form:

$$\mathbf{p}^T(\tilde{s}_1, \tilde{s}_2) = [1, \tilde{s}_1, \tilde{s}_2, \tilde{s}_1^2, \tilde{s}_2^2, \tilde{s}_1 \tilde{s}_2], \quad m = 6, \quad \tilde{s}_i = s_i - s_i^E \; ; \quad i = 1, 2 \quad (8.2)$$

The coefficients a_i and b_i are obtained by minimizing the weighted discrete L_2 norms:

$$R_u = \sum_{I=1}^{n} w_I(d) \left[\mathbf{p}^T(\mathbf{s}^I - \mathbf{s}^E)\mathbf{a} - \hat{u}_I \right]^2$$

$$R_\tau = \sum_{I=1}^{n} w_I(d) \left[\mathbf{p}^T(\mathbf{s}^I - \mathbf{s}^E)\mathbf{b} - \hat{\tau}_I \right]^2 \quad (8.3)$$

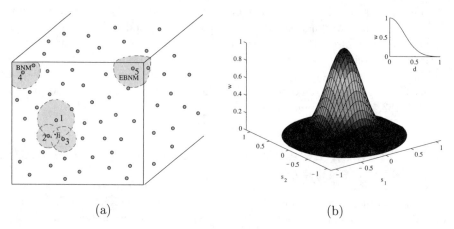

(a)   (b)

Figure 8.1: Domain of dependence and range of influence. (a) The nodes 1, 2 and 3 lie within the domain of dependence of the evaluation point E. The ranges of influence of nodes 1, 2, 3, 4 and 5 are shown as gray regions. In the standard BNM, the range of influence of a node near an edge, e.g. node 4, is truncated at the edges of a panel. In the EBNM, the range of influence can reach over to neighboring panels and contain edges and/or corners - see, e.g. node 5 (b) Gaussian weight function defined on the range of influence of a node (from [163])

where the summation is carried out over the n boundary nodes for which the weight function $w_I(d) \neq 0$ (Weight functions are defined in Section 8.3). The quantity $d = g(\mathbf{s}, \mathbf{s}^I)$ is the length of the geodesic on ∂B between $\mathbf{s}$ and $\mathbf{s}^I$. These n nodes are said to be within the domain of dependence of a point $\mathbf{s}$ (evaluation point E in Figure 8.1(a)). Also, $(s_1^I - s_1^E, s_2^I - s_2^E)$ are the local surface coordinates of the boundary nodes with respect to the evaluation point $\mathbf{s}^E = (s_1^E, s_2^E)$ and $\hat{u}_I$ and $\hat{\tau}_I$ are the approximations to the nodal values u_I and τ_I. These equations above can be rewritten in compact form as:

$$R_u = [\mathbf{P}(\mathbf{s}^I - \mathbf{s}^E)\mathbf{a} - \hat{\mathbf{u}}]^T \mathbf{W}(\mathbf{s}, \mathbf{s}^I)[\mathbf{P}(\mathbf{s}^I - \mathbf{s}^E)\mathbf{a} - \hat{\mathbf{u}}] \qquad (8.4)$$

$$R_\tau = [\mathbf{P}(\mathbf{s}^I - \mathbf{s}^E)\mathbf{b} - \hat{\boldsymbol{\tau}}]^T \mathbf{W}(\mathbf{s}, \mathbf{s}^I)[\mathbf{P}(\mathbf{s}^I - \mathbf{s}^E)\mathbf{b} - \hat{\boldsymbol{\tau}}] \qquad (8.5)$$

where $\hat{\mathbf{u}}^T = (\hat{u}_1, \hat{u}_2, \cdots, \hat{u}_n)$, $\hat{\boldsymbol{\tau}}^T = (\hat{\tau}_1, \hat{\tau}_2, \cdots, \hat{\tau}_n)$, $\mathbf{P}(\mathbf{s}^I)$ is an $n \times m$ matrix whose k_{th} row is:

$$[1, p_2(s_1^{(k)}, s_2^{(k)}), ..., p_m(s_1^{(k)}, s_2^{(k)})]$$

and $\mathbf{W}(\mathbf{s}, \mathbf{s}^I)$ is an $n \times n$ diagonal matrix with $w_{kk} = w_k(d)$ (no sum over k).

The stationarity of R_u and R_τ, with respect to $\mathbf{a}$ and $\mathbf{b}$, respectively, leads to the equations:

$$\mathbf{a}(\mathbf{s}) = \mathbf{A}^{-1}(\mathbf{s})\mathbf{B}(\mathbf{s})\hat{\mathbf{u}}, \quad \mathbf{b}(\mathbf{s}) = \mathbf{A}^{-1}(\mathbf{s})\mathbf{B}(\mathbf{s})\hat{\boldsymbol{\tau}} \qquad (8.6)$$

where

$$\mathbf{A}(\mathbf{s}) = \mathbf{P}^T(\mathbf{s}^I - \mathbf{s}^E)\mathbf{W}(\mathbf{s}, \mathbf{s}^I)\mathbf{P}(\mathbf{s}^I - \mathbf{s}^E)$$
$$\mathbf{B}(\mathbf{s}) = \mathbf{P}^T(\mathbf{s}^I - \mathbf{s}^E)\mathbf{W}(\mathbf{s}, \mathbf{s}^I) \qquad (8.7)$$

It is noted from above that the coefficients a_i and b_i turn out to be functions of $\mathbf{s}$. Substituting equations (8.6) into equations (8.1), leads to:

$$u(\mathbf{s}) = \sum_{I=1}^{n} \Phi_I(\mathbf{s})\hat{u}_I, \quad \tau(\mathbf{s}) = \sum_{I=1}^{n} \Phi_I(\mathbf{s})\hat{\tau}_I \qquad (8.8)$$

where the approximating functions Φ_I are:

$$\Phi_I(\mathbf{s}) = \sum_{j=1}^{m} p_j(\mathbf{s} - \mathbf{s}^E)(\mathbf{A}^{-1}\mathbf{B})_{jI}(\mathbf{s}) \qquad (8.9)$$

An alternative form of (8.9) is:

$$\Phi(\mathbf{s}) = \mathbf{p}^T(\mathbf{s} - \mathbf{s}^E)(\mathbf{A}^{-1}\mathbf{B})(\mathbf{s}) \qquad (8.10)$$

where $\Phi(\mathbf{s})$ is $1 \times n$.

As mentioned previously, $\hat{\mathbf{u}}$ and $\hat{\boldsymbol{\tau}}$ are approximations to their real values $\mathbf{u}$ and $\boldsymbol{\tau}$. Matrix versions of (8.8) can be written as:

$$[\mathbf{H}]\{\hat{\mathbf{u}}\} = \{\mathbf{u}\} \ , \quad [\mathbf{H}]\{\hat{\boldsymbol{\tau}}\} = \{\boldsymbol{\tau}\} \tag{8.11}$$

Equations (8.11) relate the nodal approximations of u and τ to their nodal values.

Remarks

- *Remark 1: Invertibility of* $\mathbf{A}$
  The matrix $\mathbf{A}$ is an $m \times m$ matrix, composed of the matrices $\mathbf{P}$ and $\mathbf{W}$ (equation 8.7). It needs to be invertible for the construction of the shape functions. It is also desirable that $\mathbf{A}$ be well conditioned. From a well-known fact in linear algebra about ranks of products of matrices, it is necessary that the rank of matrix $\mathbf{P}$ be m. However, if $n < m$ i.e. the number of nodes n in the domain of dependence of an evaluation point is less than the order of the polynomial basis m, then matrix $\mathbf{A}$ would be rank deficient and would become noninvertible. So, it is essential to choose the parameter which controls the range of influence of a node, namely $\hat{d}$, such that $n \geq m$. However, even if the condition $n \geq m$ is satisfied, but the n nodes in the domain of dependence of the evaluation point E lie on a straight line on the surface, then the matrix $\mathbf{A}$ becomes singular. Also, it has been observed that choosing $n \sim m$ may lead to an unacceptably large condition number of the matrix $\mathbf{A}$.

- *Remark 2: Matrix* $\mathbf{H}$
  As noted above, the matrix $\mathbf{H}$ relates the actual nodal values to their nodal approximations. It is observed through numerical experiments that the matrix $\mathbf{H}$ has m eigenvalues equal to unity. The associated m eigenvectors are described by the monomials used in the bases for constructing the approximation. Thus, when looking for solutions that *cannot* be spanned by the monomials used in the bases, the matrix $\mathbf{H}$ plays a significant role in the success of the method.

- *Remark 3: Boundary conditions*
  The $\mathbf{H}$ matrix plays a crucial role in the satisfaction of essential boundary conditions in the EFG method [108] and in the satisfaction of all boundary conditions in the BNM [107].

- *Remark 4: Definition of a panel*
  Curvilinear coordinates (s_1, s_2) are used to measure distances over curved surfaces. However, real life objects consist of piecewise smooth surfaces, referred to as *panels* in this work, and defining curvilinear coordinates across edges and corners is a formidable task. In this work, collocation nodes are placed inside panels, and, in order to circumvent the problem

of "reaching over edges," it was decided that the range of influence (ROI) of each node would be truncated at an edge or corner (Figure 8.1(a)). It will be seen through numerical experiments that restricting the range of influence of a node to the panel to which it belongs still yields acceptable results. An alternative to truncation of ROIs at edges is to use Cartesian coordinates (see Figure 8.1(a) and Section 8.4).

- *Remark 5: The nature of s_1, s_2*
  The coordinates (s_1, s_2) are the curvilinear coordinates measured along the bounding surface ∂B. These coordinates are local and *not* global. In other words, these are constructed with the origin at the evaluation point E i.e. these curvilinear coordinates will always be (0,0) at the evaluation point E. This simplifies the computation of the shape functions to some extent. Since, $(s_1 = 0, s_2 = 0)$, one has $p_1 = 1$ and $p_i = 0$ for $i = 2, \cdots, m$. This further implies that the shape function is just the first row of the matrix $\mathbf{C}$.

8.2 Surface Derivatives

Surface derivatives of the potential (or displacement) field u are required for the HBIE. These are computed as follows. With

$$\mathbf{C} = \mathbf{A}^{-1}\mathbf{B}$$

equations (8.8) and (8.9) give:

$$u(\mathbf{s}) = \sum_{I=1}^{n} \sum_{j=1}^{m} p_j(\mathbf{s} - \mathbf{s}^E) \mathbf{C}_{jI}(\mathbf{s}) \hat{u}_I \tag{8.12}$$

and the tangential derivatives of u can be written as:

$$\frac{\partial u(\mathbf{s})}{\partial s_k} = \sum_{I=1}^{n} \sum_{j=1}^{m} \left[\frac{\partial p_j}{\partial s_k}(\mathbf{s} - \mathbf{s}^E) \mathbf{C}_{jI}(\mathbf{s}) + p_j(\mathbf{s} - \mathbf{s}^E) \frac{\partial \mathbf{C}_{jI}(\mathbf{s})}{\partial s_k} \right] \hat{u}_I$$

$$k = 1, 2 \tag{8.13}$$

The derivatives of the monomials p_j can be easily computed. These are:

$$\frac{\partial \mathbf{p}^T}{\partial s_1}(s_1 - s_1^E, s_2 - s_2^E) = [0, 1, 0, 2(s_1 - s_1^E), 0, (s_2 - s_2^E)] \tag{8.14}$$

$$\frac{\partial \mathbf{p}^T}{\partial s_2}(s_1 - s_1^E, s_2 - s_2^E) = [0, 0, 1, 0, 2(s_2 - s_2^E), (s_1 - s_1^E)] \tag{8.15}$$

After some simple algebra (Chati [23]), the derivatives of the matrix $\mathbf{C}$ with respect to s_k take the form:

$$\frac{\partial \mathbf{C}(\mathbf{s})}{\partial s_k} = -\mathbf{A}^{-1}(\mathbf{s})\frac{\partial \mathbf{B}(\mathbf{s})}{\partial s_k}\mathbf{P}(\mathbf{s}^I - \mathbf{s}^E)\mathbf{A}^{-1}(\mathbf{s})\mathbf{B}(\mathbf{s}) + \mathbf{A}^{-1}(\mathbf{s})\frac{\partial \mathbf{B}(\mathbf{s})}{\partial s_k}$$

$$k = 1, 2 \qquad\qquad (8.16)$$

with

$$\frac{\partial \mathbf{B}(\mathbf{s})}{\partial s_k} = \mathbf{P}^T(\mathbf{s}^I - \mathbf{s}^E)\frac{\partial \mathbf{W}(\mathbf{s}, \mathbf{s}^I)}{\partial s_k} \qquad\qquad (8.17)$$

In deriving equation (8.16), the following identity has been used:

$$\frac{\partial \mathbf{A}^{-1}(\mathbf{s})}{\partial s_k} = -\mathbf{A}^{-1}(\mathbf{s})\frac{\partial \mathbf{A}(\mathbf{s})}{\partial s_k}\mathbf{A}^{-1}(\mathbf{s}), \qquad k = 1, 2 \qquad\qquad (8.18)$$

Tangential derivatives of the weight functions (described in Section 8.3) are easily computed (Chati [23]). The final form of the tangential derivatives of the potential (or displacement) u, at an evaluation point E, takes the form:

$$\frac{\partial u}{\partial s_k}(\mathbf{s}^E) = \sum_{I=1}^{n}\sum_{j=1}^{m}\left[\frac{\partial p_j}{\partial s_k}(0,0)\mathbf{C}_{jI}(\mathbf{s}^E)\right]\hat{u}_I$$

$$+ \sum_{I=1}^{n}\sum_{j=1}^{m}\left[p_j(0,0)\left[\mathbf{A}^{-1}(\mathbf{s}^E)\frac{\partial \mathbf{B}}{\partial s_k}(\mathbf{s}^E)\left(\mathbf{I} - \mathbf{P}(\mathbf{s}^I - \mathbf{s}^E)\mathbf{A}^{-1}(\mathbf{s}^E)\mathbf{B}(\mathbf{s}^E)\right)\right]_{jI}\right]\hat{u}_I$$

$$(8.19)$$

with $k = 1, 2$. In the above equation, $\mathbf{I}$ is the identity matrix.

One also needs the spatial gradient of the function u in order to solve the HBIE. For problems in potential theory, this is easily obtained from its tangential and normal derivatives, i.e. $\partial u/\partial s_k$ and $\partial u/\partial n$ (see (1.14)). For elasticity problems, however, one must also use Hooke's law at a point on the surface ∂B. Details of this procedure are given in Chati et al. [27] and in Chapter 1 of this book.

Equation (8.19) can be rewritten in compact form as:

$$\frac{\partial u}{\partial s_k}(\mathbf{s}^E) = \sum_{I=1}^{n}\Psi_I^{(k)}(\mathbf{s}^E)\hat{u}_I \quad ; \qquad k = 1, 2 \qquad\qquad (8.20)$$

where the approximating functions $\Psi_I^{(k)}$ are:

$$\Psi_I^{(k)}(\mathbf{s}^E) = \sum_{j=1}^{m}\left[\frac{\partial p_j}{\partial s_k}(0,0)\mathbf{C}_{jI}(\mathbf{s}^E)\right]$$

$$+ \sum_{j=1}^{m} \left[p_j(0,0) \left[\mathbf{A}^{-1}(\mathbf{s}^E) \frac{\partial \mathbf{B}}{\partial s_k}(\mathbf{s}^E) \left(\mathbf{I} - \mathbf{P}(\mathbf{s}^I - \mathbf{s}^E) \mathbf{A}^{-1}(\mathbf{s}^E) \mathbf{B}(\mathbf{s}^E) \right) \right]_{jI} \right]$$

$$(8.21)$$

8.3   Weight Functions

The basic idea behind the choice of a weight function is that its value should decrease with distance from a node and that it should have compact support so that the region of influence of a node is of finite extent (Figure 8.1(b)). Possible choices of weight functions [26] are:

- Gaussian (referred to as - WFA) :

$$w_I(d) = \begin{cases} e^{-(d/d_I)^2} & \text{for } d \leq d_I \\ 0 & \text{for } d > d_I \end{cases} \qquad (8.22)$$

- Exponential (referred to as - WFB) :

$$w_I(d) = \begin{cases} \dfrac{e^{-(d/c)^2} - e^{(d_I/c)^2}}{1 - e^{(d_I/c)^2}} & \text{for } d \leq d_I \\ 0 & \text{for } d > d_I \end{cases} \qquad (8.23)$$

- Cubic Spline (referred to as - WFC) :

$$w_I(d) = \begin{cases} 2/3 - 4(\hat{d})^2 + 4(\hat{d})^3 & \text{for } \hat{d} \leq 1/2 \\ 4/3 - 4(\hat{d}) + 4(\hat{d})^2 - (4/3)(\hat{d})^3 & \text{for } 1/2 < \hat{d} \leq 1 \\ 0 & \text{for } d > 1 \end{cases} \qquad (8.24)$$

- Quartic Spline (referred to as - WFD) :

$$w_I(d) = \begin{cases} 1 - 6(\hat{d})^2 + 8(\hat{d})^3 - 3(\hat{d})^4 & \text{for } \hat{d} \leq 1 \\ 0 & \text{for } \hat{d} > 1 \end{cases} \qquad (8.25)$$

where $\hat{d} = d/d_I$ and c is a constant.

Here $d = g(\mathbf{s}, \mathbf{s^I})$ is the *minimum distance*, measured on the surface ∂B, (i.e. the geodesic) between a point $\mathbf{s}$ and the collocation node I. In the research performed to date, the region of influence of a node has been truncated at the edge of a panel (Figure 8.1(a)) so that geodesics, and their derivatives (for use in equation (8.17)), need only be computed on piecewise smooth surfaces. Finally, the quantities d_I determine the extent of the region of influence (the compact support) of node I. They can be made globally uniform, or can be adjusted such that approximately the same number of nodes get included in the region of influence of any given node I or in the domain of dependence of a given evaluation point E. Such ideas have been successfully implemented in Chati and Mukherjee [26] and Chati et al. [25].

8.4 Use of Cartesian Coordinates

One of the drawbacks of using curvilinear surface coordinates as described above in Section 8.1 is the need to truncate the range of influence of a node at an edge or corner (see Remark 4 in Section 8.1). Another is the need to compute geodesics on general surfaces. The more straightforward approach, namely the use of Cartesian coordinates, suffers from the disadvantage that the matrix $\mathbf{A}$ defined in equation (8.7) becomes singular if all the nodes in the domain of dependence of an evaluation point lie on a plane (see [117, 79]). Li and Aluru have suggested, in two recent papers [79, 80], ways to use Cartesian coordinates in a modified version of the BNM which they call the boundary cloud method. (The acronym BCLM is used for the boundary cloud method in this book). Nice results for 2-D problems in potential theory are given in [79, 80]. This idea is discussed below for 3-D problems in potential theory with linear approximants. Extension to 3-D elasticity is relatively straightforward.

8.4.1 Hermite Type Approximation

For the 3-D Laplace equation (see [79] for the 2-D case), one writes:

$$u(\mathbf{x}) = \mathbf{p}^T(\mathbf{x})\mathbf{a}, \qquad \tau(\mathbf{x}) = \frac{\partial \mathbf{p}^T}{\partial n}(\mathbf{x})\mathbf{a} \qquad (8.26)$$

Collocation is not allowed at a point on an edge or a corner. For a linear approximation:

$$\mathbf{p}^T(\mathbf{x}) = [1, x_1, x_2, x_3]$$
$$\frac{\partial \mathbf{p}^T}{\partial n}(\mathbf{x}) = \left[0, \frac{\partial x_1}{\partial n}, \frac{\partial x_2}{\partial n}, \frac{\partial x_3}{\partial n}\right] = [0, n_1, n_2, n_3] \qquad (8.27)$$

where $\mathbf{n}$ is the unit outward normal at a boundary point.

The coefficients a_i are obtained by minimizing the weighted discrete L_2 norm:

$$J = \sum_{I=1}^{n} w_I(\mathbf{x}^t, \mathbf{x}^I)\left[\mathbf{p}^T(\mathbf{x}^I)\mathbf{a} - \hat{u}_I\right]^2 + \sum_{I=1}^{n} w_I(\mathbf{x}^t, \mathbf{x}^I)\left[\frac{\partial \mathbf{p}^T}{\partial n}(\mathbf{x}^I)\mathbf{a} - \hat{\tau}_I\right]^2 \quad (8.28)$$

In [79], where a fixed least squares approach is adopted, $\mathbf{x}^t$ are the coordinates of a fixed point inside a cloud (chosen to be the center of a cloud), and $\mathbf{x}^I$, as before, are the coordinates of node I. This time, the Euclidean distance between $\mathbf{x}^t$ and $\mathbf{x}^I$ is used in the weight function. The weight functions are piecewise constant, i.e. they are constant within each cloud but vary from cloud to cloud.

The stationarity of J with respect to $\mathbf{a}$ leads to the equations:

$$u(\mathbf{x}) = \sum_{I=1}^{n} M_I(\mathbf{x})\hat{u}_I + \sum_{I=1}^{n} N_I(\mathbf{x})\hat{\tau}_I \tag{8.29}$$

$$\tau(\mathbf{x}) = \sum_{I=1}^{n} S_I(\mathbf{x})\hat{u}_I + \sum_{I=1}^{n} T_I(\mathbf{x})\hat{\tau}_I \tag{8.30}$$

where:

$$\mathbf{M}(\mathbf{x}) = \mathbf{p}^T(\mathbf{x})(\mathbf{A}^{-1}\mathbf{B}), \qquad \mathbf{N} = \mathbf{p}^T(\mathbf{x})(\mathbf{A}^{-1}\mathbf{D}) \tag{8.31}$$

with:

$$\mathbf{S}(\mathbf{x}) = \frac{\partial \mathbf{p}^T}{\partial n}(\mathbf{x})(\mathbf{A}^{-1}\mathbf{B}), \qquad \mathbf{T}(\mathbf{x}) = \frac{\partial \mathbf{p}^T}{\partial n}(\mathbf{x})(\mathbf{A}^{-1}\mathbf{D}) \tag{8.32}$$

with:

$$\mathbf{A}(\mathbf{x}) = \mathbf{P}^T(\mathbf{x}^I)\mathbf{W}(\mathbf{x}^t,\mathbf{x}^I)\mathbf{P}(\mathbf{x}^I) + \frac{\partial \mathbf{P}^T}{\partial n}(\mathbf{x}^I)\mathbf{W}(\mathbf{x}^t,\mathbf{x}^I)\frac{\partial \mathbf{P}}{\partial n}(\mathbf{x}^I) \tag{8.33}$$

$$\mathbf{B}(\mathbf{x}) = \mathbf{P}^T(\mathbf{x}^I)\mathbf{W}(\mathbf{x}^t,\mathbf{x}^I) \tag{8.34}$$

$$\mathbf{D}(\mathbf{x}) = \frac{\partial \mathbf{P}^T}{\partial n}(\mathbf{x}^I)\mathbf{W}(\mathbf{x}^t,\mathbf{x}^I) \tag{8.35}$$

It is proved in [79] that the matrix $\mathbf{A}$ in (8.33) is nonsingular.

8.4.2   Variable Basis Approximation

Li and Aluru [80] present a variable basis approach for solving the 2-D Laplace's equation with the boundary cloud method (BCLM). This is an elegant approach in which reduced bases are appropriately employed in order to avoid singularity of the matrix $\mathbf{A} = \mathbf{P}^T\mathbf{W}\mathbf{P}$. A disadvantage of this approach, as well as that of the standard BNM, is that discontinuities in the normal derivative of the potential function, across edges and corners, are not addressed properly - normal derivatives are modeled with continuous approximants, even across edges and corners. Very recently, Telukunta and Mukherjee [163, 164] have combined the advantages of the variable basis approach [80], together with allowing discontinuities in $\tau = \frac{\partial u}{\partial n}$, in a new approach called the extended boundary node method (EBNM). The EBNM, for 2-D and 3-D potential theory, is described next.

It is important to mention here that the fixed least squares approach has been adopted by Li and Aluru in [79, 80] while the variable basis approach adopted in [163, 164] uses moving least squares for both the BCLM and the EBNM. The title "BCM" used in figures depicting numerical results in [163] refer to a variable basis BCLM with moving least squares. These results have been obtained from a fresh implementation of the equations presented in [80], adapted to the moving least squares formulation. The quantity d in the argument of a weight function w_I in Section 8.3 is, in general, still the geodesic between $\mathbf{x}$ and $\mathbf{x}^I$ in the EBNM. As mentioned before, d is taken as the Euclidean distance between $\mathbf{x}^t$ (a fixed point inside a cloud) and $\mathbf{x}^I$ in the work of Li and Aluru [79, 80] that adopts a fixed least squares approach.

The first step is to distinguish between singular and nonsingular clouds (DODs and ROIs are sometimes called clouds in this work - the term is taken from the work of Li and Aluru [79, 80]). A **straight cloud** (for 2-D problems) is one in which the nodes lie on a straight line. Similarly, a **flat cloud** (for 3-D problems) is one in which the nodes lie on a plane. A **curved cloud** is a smooth curve in 2-D and a smooth surface in 3-D problems. Finally, a **broken cloud** contains at least one corner in 2-D and at least one edge or corner in 3-D problems. *The approximants for u are identical in the variable basis BCLM [80] and in the EBNM [163]. They are, however, different for τ - the approximants for τ are continuous (same as those for u) in [80] but allow for jumps in τ across corners and edges in the EBNM [163]. The specific approximants for the EBNM are given below.*

The starting point is to write the approximations (8.1) in terms of Cartesian coordinates $\mathbf{x}$:

$$u(\mathbf{x}) = \mathbf{p}^T(\mathbf{x})\mathbf{a} , \qquad \tau(\mathbf{x}) = \mathbf{q}^T(\mathbf{x})\mathbf{b} \qquad (8.36)$$

A basis (i.e. the functions in $\mathbf{p}$ or $\mathbf{q}$ in (8.36)) for a cloud must satisfy two competing requirements - it must be broad enough to include all cases, yet it must be narrow enough such that the matrices $\mathbf{A} = \mathbf{P}^T\mathbf{W}\mathbf{P}$ and $\mathbf{C} = \mathbf{Q}^T\mathbf{W}\mathbf{Q}$ (see (8.46 - 8.47)) are nonsingular.

It is noted here that the smoothness of the final approximating functions for u and τ (see equation (8.44)), at a regular point on the bounding surface of a body, depends on the choice of the weight function w_I [10].

8.4.2.1 Two-dimensional problems

Bases for u and τ. The following bases are used for the various cases listed below.

> *Straight cloud:* $x_1 = c_1$ Basis $[1, x_2]$ for u and τ

All other straight clouds: Basis $[1, x_1]$ for u and τ
All curved clouds: Basis $[1, x_1, x_2]$ for u and τ

$$Broken\ cloud:\ \text{Basis}\ \begin{cases} [1, x_1, x_2] \quad \text{for } u \\[2mm] [n_1, n_1 x_2, n_2, n_2 x_1] \quad \text{for } \tau \quad \text{with both segments} \\ \hphantom{[n_1, n_1 x_2, n_2, n_2 x_1] \quad \text{for } \tau \quad} \text{straight} \\[2mm] [n_1, n_1 x_1, n_1 x_2, n_2, n_2 x_1, n_2 x_2] \quad \text{for } \tau \quad \text{with at least} \\ \hphantom{[n_1, n_1 x_1, n_1 x_2, n_2, n_2 x_1, n_2 x_2] \quad \text{for } \tau \quad} \text{one segment curved} \end{cases}$$

Explanation for choice of basis for u on a broken cloud. It is assumed that $u(x_1, x_2) \in C^\infty$ in B. Let a corner C on ∂B have coordinates (x_{10}, x_{20}) which, for simplicity, is written as (x_0, y_0). A Taylor series expansion for u about (x_0, y_0) is of the form:

$$u(x, y) = u(x_0, y_0) + u_x(x_0, y_0)(x - x_0) + u_y(x_0, y_0)(y - y_0) + h.o.t. \quad (8.37)$$

The linear approximation of u about C is of the form:

$$l(x, y) = a_0 + a_1 x + a_2 y \quad (8.38)$$

which justifies the chosen basis.

Explanation for choice of basis for τ on a broken cloud. A general curved cloud segment has the equation $f(x_1, x_2) = 0$. Assuming $\nabla u \in C^\infty$ in B, one has:

$$\tau = u_{,1}(x_1, x_2)n_1 + u_{,2}(x_1, x_2)n_2 \quad (8.39)$$

One can, therefore, use the basis $n_1[1, x_1, x_2] \cup n_2[1, x_1, x_2]$ on a general curved cloud segment.

Special cases of straight segments of a broken cloud are as follows.

Straight line: $n_1 x_1 = c_1$ Basis $n_1[1, x_2]$

Straight line: $n_2 x_2 = c_2$ Basis $n_2[1, x_1]$

Straight line: $n_1 x_1 + n_2 x_2 = c_3$ Basis $n_1[1, x_1 \text{ or } x_2] \cup n_2[1, x_1 \text{ or } x_2]$.

It is clear that the recommended reduced basis for τ covers all these cases.

Invertibility of $\mathbf{C} = \mathbf{Q}^T \mathbf{W} \mathbf{Q}$ for τ on a broken cloud.

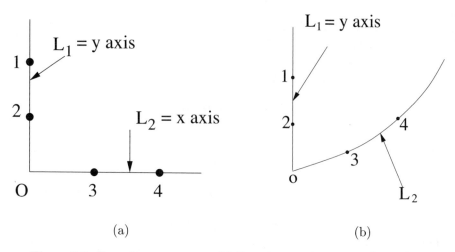

(a)　　　　　　　　　　　　　　　　　(b)

Figure 8.2: Boundary segments of 2-D regions with corners (from [163])

Straight segments with reduced basis.　Let $(x_1, x_2) \rightarrow (x, y)$ and $[n_1, n_2] \rightarrow [p, q]$. Referring to Figure 8.2(a) (the worst case scenario), it is easy to show that:

$$x_1 = x_2 = 0, \quad y_3 = y_4 = 0$$
$$q_1 = q_2 = 0, \quad p_1 = p_2 = -1$$
$$p_3 = p_4 = 0, \quad q_3 = q_4 = -1 \tag{8.40}$$

With the reduced basis $[p, py, q, qx]$, one has:

$$[C] = \begin{bmatrix} \sum_{L_1} w_i & \sum_{L_1} w_i y_i & 0 & 0 \\ \sum_{L_1} w_i y_i & \sum_{L_1} w_i y_i^2 & 0 & 0 \\ 0 & 0 & \sum_{L_2} w_i & \sum_{L_2} w_i x_i \\ 0 & 0 & \sum_{L_2} w_i x_i & \sum_{L_2} w_i x_i^2 \end{bmatrix} \tag{8.41}$$

This matrix is nonsingular provided that each cloud segment contains at least two nodes. Clouds must be chosen to fulfill this requirement.

Straight and curved segments with full basis.　Figure 8.2(b) shows a broken cloud with one straight and one curved segment. (Without loss of generality, L_1 is chosen to be part of the y axis). For this case, using the full basis $[p, px, py, q, qx, qy]$, one gets the matrix $[C]$ that is displayed on the next page.

This matrix is nonsingular if the segment L_2 is curved. If, however, L_2 is straight, one has $q_i = \beta p_i$ for all nodes on L_2, with β a constant. In this case, *row 5 = $\beta \times$ row 2* and the matrix becomes singular!

$$
[C] = \begin{bmatrix}
\Sigma_{L_1} w_i + \Sigma_{L_2} p_i^2 w_i & \Sigma_{L_2} p_i^2 w_i x_i & \Sigma_{L_1} w_i y_i + \Sigma_{L_2} p_i^2 w_i y_i & \Sigma_{L_2} p_i q_i w_i & \Sigma_{L_2} p_i q_i w_i x_i & \Sigma_{L_2} p_i q_i w_i y_i \\[4pt]
 & \Sigma_{L_2} p_i^2 w_i x_i^2 & \Sigma_{L_2} p_i^2 w_i x_i y_i & \Sigma_{L_2} p_i q_i w_i x_i & \Sigma_{L_2} p_i q_i w_i x_i^2 & \Sigma_{L_2} p_i q_i w_i x_i y_i \\[4pt]
 & & \Sigma_{L_1} w_i y_i^2 + \Sigma_{L_2} p_i^2 w_i y_i^2 & \Sigma_{L_2} p_i q_i w_i y_i & \Sigma_{L_2} p_i q_i w_i x_i y_i & \Sigma_{L_2} p_i q_i w_i y_i^2 \\[4pt]
 & & & \Sigma_{L_2} q_i^2 w_i & \Sigma_{L_2} q_i^2 w_i x_i & \Sigma_{L_2} q_i^2 w_i y_i \\[4pt]
 & \text{symmetric} & & & \Sigma_{L_2} q_i^2 w_i x_i^2 & \Sigma_{L_2} q_i^2 w_i x_i y_i \\[4pt]
 & & & & & \Sigma_{L_2} q_i^2 w_i y_i^2
\end{bmatrix}
$$

The matrix [C]

8.4.2.2 Three-dimensional problems

First, a word of caution. In order for the matrices $\mathbf{A}$ and $\mathbf{C}$ (see (8.46 - 8.47))
to be invertible, a flat cloud (or a flat segment of a broken cloud) must always
contain three or more points, and all the points on it must not lie on a straight
line [164].

Bases for u and τ. The following bases are used for the various cases listed
below.

Flat cloud: $\left\{ \begin{array}{llll} x_1 = c_1 & \text{or} \quad n_1x_1 + n_2x_2 = c_4 & \text{Basis} \ [1, x_2, x_3] & \text{for } u \text{ and } \tau \\ x_2 = c_2 & \text{or} \quad n_2x_2 + n_3x_3 = c_5 & \text{Basis} \ [1, x_3, x_1] & \text{for } u \text{ and } \tau \end{array} \right.$

$$\begin{array}{lll} \textit{All other flat clouds:} & \text{Basis} \quad [1, x_1, x_2] & \text{for } u \text{ and } \tau \\ \textit{All curved clouds:} & \text{Basis} \quad [1, x_1, x_2, x_3] & \text{for } u \text{ and } \tau \end{array}$$

$$\textit{Broken cloud:} \quad \text{Basis} \left\{ \begin{array}{l} [1, x_1, x_2, x_3] \quad \text{for } u \\ \text{see below for } \tau \end{array} \right.$$

Explanation for choice of basis for u on a broken cloud. The arguments
given for the 2-D case can be easily extended to the 3-D case as well.

Choice of basis for τ on a broken cloud. A general curved cloud segment
has the equation $f(x_1, x_2, x_3) = 0$. Assuming $\boldsymbol{\nabla} u \in C^\infty$ in B, one has:

$$\tau = u_{,1}(x_1, x_2, x_3)n_1 + u_{,2}(x_1, x_2, x_3)n_2 + u_{,3}(x_1, x_2, x_3)n_3 \qquad (8.42)$$

One can, therefore, use the basis $\quad n_1[1, x_1, x_2, x_3] \ \cup \ n_2[1, x_1, x_2, x_3] \ \cup$
$n_3[1, x_1, x_2, x_3]$ on a *general curved cloud segment*.

Special cases of flat segments in a broken cloud

Plane P_1: $n_1x_1 = c_1$ Basis $n_1[1, x_2, x_3]$

Plane P_2: $n_2x_2 = c_2$ Basis $n_2[1, x_3, x_1]$

Plane P_3: $n_3x_3 = c_3$ Basis $n_3[1, x_1, x_2]$

Plane P_4: $n_1x_1 + n_2x_2 = c_4$ Basis $n_1[1, x_1 \text{ or } x_2, x_3] \cup n_2[1, x_1 \text{ or } x_2, x_3]$

Plane P_5: $n_2x_2 + n_3x_3 = c_5$ Basis $n_2[1, x_1, x_2 \text{ or } x_3] \cup n_3[1, x_1, x_2 \text{ or } x_3]$

Plane P_6: $n_3x_3 + n_1x_1 = c_6$ Basis $n_3[1, x_1 \text{ or } x_3, x_2] \cup n_1[1, x_1 \text{ or } x_3, x_2]$

Plane P_7: $\quad n_1 x_1 + n_2 x_2 + n_3 x_3 = c_7 \quad$ Basis $\quad n_1[1, \text{ two of } (x_1, x_2, x_3)] \cup$
$$n_2[1, \text{ two of } (x_1, x_2, x_3)] \cup n_3[1, \text{ two of } (x_1, x_2, x_3)]$$

On a broken cloud, one must use a basis that is a union of the bases for its segments.

For a union of two flat segments, one has, for example:

$P_1 \cup P_2$: $\quad$ Basis $\quad [n_1, n_1 x_2, n_1 x_3, n_2, n_2 x_3, n_2 x_1]$

$P_1 \cup P_4$: $\quad$ Basis $\quad [n_1, n_1 x_2, n_1 x_3, n_2, n_2 x_3, n_2 x_1]$

$P_1 \cup P_5$: $\quad$ Basis $\quad [n_1, n_1 x_2, n_1 x_3, n_2, n_2 x_3, n_2 x_1, n_3, n_3 x_1, n_3 x_2]$

$P_1 \cup P_7$: $\quad$ Basis $\quad [n_1, n_1 x_2, n_1 x_3, n_2, n_2 x_3, n_2 x_1, n_3, n_3 x_1, n_3 x_2]$

For a union of three flat segments, one has, for example:

$P_1 \cup P_2 \cup P_3$: $\quad$ Basis $\quad [n_1, n_1 x_2, n_1 x_3, n_2, n_2 x_3, n_2 x_1, n_3, n_3 x_1, n_3 x_2]$

Special cases of curved segments in a broken cloud

S_1: $\quad f_1(x_1, x_2) = 0 \quad$ Basis $\quad [n_1, n_1 x_1, n_1 x_2, n_1 x_3, n_2, n_2 x_1, n_2 x_2, n_2 x_3]$

S_2: $\quad f_2(x_2, x_3) = 0 \quad$ Basis $\quad [n_2, n_2 x_1, n_2 x_2, n_2 x_3, n_3, n_3 x_1, n_3 x_2, n_3 x_3]$

S_3: $\quad f_3(x_3, x_1) = 0 \quad$ Basis $\quad [n_1, n_1 x_1, n_1 x_2, n_1 x_3, n_3, n_3 x_1, n_3 x_2, n_3 x_3]$

S_4: $\quad f_4(x_1, x_2, x_3) = 0$
$\quad$ Basis $\quad [n_1, n_1 x_1, n_1 x_2, n_1 x_3, n_2, n_2 x_1, n_2 x_2, n_2 x_3, n_3, n_3 x_1, n_3 x_2, n_3 x_3]$

In the above, f_k, $k = 1, 2, 3, 4$, are nonlinear functions of their arguments.

On a broken cloud, one must use a basis that is a union of the bases for its segments.

For example:

$P_1 \cup S_1$: $\quad$ Basis $\quad [n_1, n_1 x_1, n_1 x_2, n_1 x_3, n_2, n_2 x_1, n_2 x_2, n_2 x_3]$

$P_1 \cup S_2$: $\quad$ Basis $\quad [n_1, n_1 x_2, n_1 x_3, n_2, n_2 x_1, n_2 x_2, n_2 x_3, n_3, n_3 x_1, n_3 x_2, n_3 x_3]$

Note that the case $P_1 \cup S_1$ is a 3-D version of the 2-D example in Figure 8.2(b).

Invertibility of $\mathbf{C} = \mathbf{Q}^T\mathbf{W}\mathbf{Q}$ for τ on a broken cloud with flat segments. The matrix $\mathbf{C}$ becomes singular if and only if all the points on any flat cloud segment are colinear. A proof of this fact is available in [164].

8.4.2.3   Determination of approximating functions for u and τ

Next, analogous to (8.3), let:

$$S_u = \sum_{I=1}^{n} w_I(d) \left[\mathbf{p}^T(\mathbf{x}^I)\mathbf{a} - \hat{u}_I\right]^2 , \quad S_\tau = \sum_{I=1}^{n} w_I(d) \left[\mathbf{q}^T(\mathbf{x}^I)\mathbf{b} - \hat{\tau}_I\right]^2 \quad (8.43)$$

Following the same steps as in Section 8.1, one finally gets:

$$u(\mathbf{x}) = \sum_{I=1}^{n} \alpha_I(\mathbf{x})\hat{u}_I, \quad \tau(\mathbf{x}) = \sum_{I=1}^{n} \beta_I(\mathbf{x})\hat{\tau}_I \quad (8.44)$$

where:

$$\boldsymbol{\alpha}(\mathbf{x}) = \mathbf{p}^T(\mathbf{x})(\mathbf{A}^{-1}\mathbf{B})(\mathbf{x}), \quad \boldsymbol{\beta}(\mathbf{x}) = \mathbf{q}^T(\mathbf{x})(\mathbf{C}^{-1}\mathbf{D})(\mathbf{x}) \quad (8.45)$$

with:

$$\mathbf{A}(\mathbf{x}) = \mathbf{P}^T(\mathbf{x}^I)\mathbf{W}(\mathbf{x},\mathbf{x}^I)\mathbf{P}(\mathbf{x}^I), \quad \mathbf{B}(\mathbf{x}) = \mathbf{P}^T(\mathbf{x}^I)\mathbf{W}(\mathbf{x},\mathbf{x}^I) \quad (8.46)$$

$$\mathbf{C}(\mathbf{x}) = \mathbf{Q}^T(\mathbf{x}^I)\mathbf{W}(\mathbf{x},\mathbf{x}^I)\mathbf{Q}(\mathbf{x}^I), \quad \mathbf{D}(\mathbf{x}) = \mathbf{Q}^T(\mathbf{x}^I)\mathbf{W}(\mathbf{x},\mathbf{x}^I) \quad (8.47)$$

In the above, $\mathbf{P}(\mathbf{x}^I)$ is an $n \times m$ matrix whose k_{th} row is:

$$[p_1(\mathbf{x}^k), p_2(\mathbf{x}^k), ..., p_m(\mathbf{x}^k)]$$

similarly for $\mathbf{Q}(\mathbf{x}^I)$; and $\mathbf{W}(\mathbf{x},\mathbf{x}^I)$ is an $n \times n$ diagonal matrix with $w_{kk} = w_k(d)$ (no sum over k).

Numerical results from the EBNM, for 2-D and 3-D problems in potential theory [163, 164], are most encouraging.

It is important to mention here that the rest of this part of this book presents the boundary node method with curvilinear surface coordinates (s_1, s_2). Use of Cartesian coordinates in the 3-D BNM is an interesting subject of continuing research.

Chapter 9

POTENTIAL THEORY AND ELASTICITY

This chapter presents a procedure for coupling of 3-D singular and hypersingular integral equations (BIE and HBIE - see Chapter 1), with moving least squares (MLS) interpolants (see Chapter 8), to obtain the singular and hypersingular boundary node method (BNM and HBNM) equations. Numerical results for selected examples follow. Potential theory is presented first; then linear elasticity.

9.1 Potential Theory in Three Dimensions

9.1.1 BNM: Coupling of BIE with MLS Approximants

9.1.1.1 Coupled equations

The bounding surface ∂B of a body B is partitioned into N_c cells ∂B_k and MLS approximations for the functions u and τ (8.8) are used in the usual BIE for 3-D potential theory (1.7). The result is:

$$
0 = \sum_{k=1}^{N_c} \int_{\partial B_k} \left[G(\mathbf{x}, \mathbf{y}) \sum_{I=1}^{n_y} \Phi_I(\mathbf{y}) \hat{\tau}_I - F(\mathbf{x}, \mathbf{y}) \left\{ \sum_{I=1}^{n_y} \Phi_I(\mathbf{y}) \hat{u}_I - \sum_{I=1}^{n_x} \Phi_I(\mathbf{x}) \hat{u}_I \right\} \right] dS(\mathbf{y})
$$

$$(9.1)$$

where $\Phi_I(\mathbf{x})$ and $\Phi_I(\mathbf{y})$ are the contributions from the I_{th} node to the collocation point $\mathbf{x}$ and field point $\mathbf{y}$ respectively. Also, n_y nodes lie in the domain of dependence of the field point $\mathbf{y}$ and n_x nodes lie in the domain of dependence of the source point $\mathbf{x}$. When $\mathbf{x}$ and $\mathbf{y}$ belong to the same cell, the cell is treated as a singular cell and the special techniques described in the next subsection

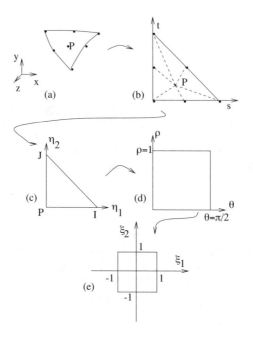

Figure 9.1: Mapping scheme for weakly singular integrals (from [26])

are used to carry out the integrations. Otherwise, regular Gaussian integration is used.

9.1.1.2   Weakly singular integration scheme

As mentioned above, regular Gaussian integration can be used as long as the source and field points are on different cells. However, the kernels become singular as the source and collocation points approach each other, i.e. as $\mathbf{y} \to \mathbf{x}$. The kernel $F(\mathbf{x}, \mathbf{y})$ is strongly singular $(\mathcal{O}(1/r^2))$ while the kernel $G(\mathbf{x}, \mathbf{y})$ is weakly singular $(O(1/r))$ as $r \to 0$ (here r is the Euclidean distance between $\mathbf{x}$ and $\mathbf{y}$). The strongly singular integral in (9.1) is regularized in the usual way by multiplying the kernel $F(\mathbf{x}, \mathbf{y})$ by the $\mathcal{O}(r)$ function $u(\mathbf{y}) - u(\mathbf{x})$. Various methods have been proposed in the literature to handle weakly singular integrals arising in the BEM. The method suggested by Nagarajan and Mukherjee [114] has been used to carry out the weakly singular integration in the BNM. The details follow.

Consider evaluating the integral with $G(\mathbf{x}, \mathbf{y})$ in (9.1) over a (in general curved) surface cell as shown in Figure 9.1(a). This integral can be represented as:

$$I = \int_{\partial B} O(1/r) \, dS_Q \tag{9.2}$$

The cell shown contains the source point $\mathbf{x}$, so that the field point $\mathbf{y}$ could coincide with the collocation point $\mathbf{x}$. In this example, Quadratic (T6) triangles are used to describe the geometry of the bounding surface. The method described here, however, can be easily extended for various other kinds of geometric interpolations. First, the cell is mapped into the parent space (Figure 9.1(b)) using the well-known shape functions for T6 triangles. This involves a Jacobian and the integral I takes the form:

$$I = \int_{t=0}^{t=1} \int_{s=0}^{s=1-t} O(1/r) J_1 \, ds \, dt \tag{9.3}$$

Now, in the parent space the triangle is divided into six pieces. Each individual triangle is mapped into the parametric ($\eta_1 - \eta_2$) space (Figure 9.1(c)) using the mapping for linear (T3) triangles. The integral I can now be written as:

$$I = \sum_{i=1}^{6} \int_{\eta_2=0}^{\eta_2=1} \int_{\eta_1=0}^{\eta_1=1-\eta_2} O(1/r) J_1 J_2^{(i)} \, d\eta_1 \, d\eta_2 \tag{9.4}$$

where $J_2^{(i)}$ is the Jacobian for each triangle. Now, consider the mapping [114]:

$$\eta_1 = \rho \cos^2 \theta \quad , \quad \eta_2 = \rho \sin^2 \theta \tag{9.5}$$

which maps the flat triangle from the $\eta_1 - \eta_2$ coordinate system into a rectangle in the $\rho - \theta$ space (Figure 9.1(d)). The integral I now takes the form:

$$I = \sum_{i=1}^{6} \int_{\theta=0}^{\theta=\pi/2} \int_{\rho=0}^{\rho=1} O(1/r) J_1 J_2^{(i)} \rho \sin \theta \, d\rho \, d\theta \tag{9.6}$$

As ρ is a measure of the distance between the source point and the field point, the integral I is now regularized. In other words, the ρ in the numerator cancels the $\mathcal{O}(1/r)$ singularity. Now, to evaluate the integral I in equation (9.6), regular Gaussian integration can be used. The final mapping involves the use of quadratic (Q4) shape functions to map the rectangle from the $\rho - \theta$ space into the standard square in $\xi_1 - \xi_2$ space (Figure 9.1(e)). The final form of the integral I is:

$$I = \sum_{i=1}^{6} \int_{\xi_2=-1}^{\xi_2=1} \int_{\xi_1=-1}^{\xi_1=1} O(1/r) J_1 J_2^{(i)} J_3 \rho \sin \theta \, d\xi_1 \, d\xi_2 \tag{9.7}$$

where J_3 is the Jacobian of the final transformation. Finally, regular Gaussian integration can be used to evaluate the above integral I.

9.1.1.3 Discretized equations and boundary conditions

The discretized assembled form of equation (9.1) becomes:

$$[\mathbf{K_1}]\{\hat{\mathbf{u}}\} + [\mathbf{K_2}]\{\hat{\boldsymbol{\tau}}\} = \{\mathbf{0}\}, \qquad N_B \text{ equations} \qquad (9.8)$$

where $\{\hat{\mathbf{u}}\}$ and $\{\hat{\boldsymbol{\tau}}\}$ contain the approximations to the nodal values of u and τ at the N_B boundary nodes.

Satisfaction of boundary conditions is carried out next. A special procedure is needed here since, due to the lack of the delta function property of MLS approximants, (9.1) contains approximations to the nodal values of the primary variables rather than the variables themselves [107, 108]. It is assumed that in a general mixed boundary value problem, either u or τ is prescribed at each boundary node. Let the vector $\{\bar{\mathbf{y}}\}$ contain the prescribed boundary conditions and $\{\mathbf{x}\}$ contain the rest. Each of these vectors is of length N_B. Also let $\{\hat{\mathbf{y}}\}$ and $\{\hat{\mathbf{x}}\}$ be their corresponding approximations. Finally, let $\{\hat{\mathbf{z}}\} = (\{\hat{\mathbf{u}}\} \cup \{\hat{\boldsymbol{\tau}}\})$.

Equation (9.8) is now written as:

$$[\mathbf{M}]\{\hat{\mathbf{z}}\} = \{\mathbf{0}\}, \qquad N_B \text{ equations} \qquad (9.9)$$

Referring to equations (8.11), the nodes with prescribed quantities are considered first. This gives rise to a system of equations of the form:

$$[\mathbf{H_1}]\{\hat{\mathbf{z}}\} = \{\bar{\mathbf{y}}\}, \qquad N_B \text{ equations} \qquad (9.10)$$

Equations (9.9) and (9.10) are now solved together for $2N_B$ unknowns $\hat{z}_k$. Finally, (as a postprocessing step) consideration of the rest of the boundary nodes (those without prescribed boundary conditions) results in the equations:

$$\{\mathbf{x}\} = [\mathbf{H_2}]\{\hat{\mathbf{z}}\}, \qquad N_B \text{ equations} \qquad (9.11)$$

which yield the required boundary values x_k.

9.1.1.4 Potential and potential gradient at internal points

The potential at a point $\boldsymbol{\xi}$ inside the body B is obtained from equation (1.2) while the potential gradient is obtained from (1.9); together with (8.8) in each case. It is well known that direct use of these equations can yield poor results for the potential and terrible results for the potential gradient at internal points that are close to the boundary of the body. This phenomenon, sometimes referred to as the *boundary layer effect*, is a consequence of the fact that the kernels $G(\boldsymbol{\xi}, \mathbf{y})$ and $F(\boldsymbol{\xi}, \mathbf{y})$ become *nearly singular* and *nearly hypersingular*, respectively, as $\boldsymbol{\xi} \to \mathbf{y}$. A possible remedy for the BEM for elasticity problems is discussed in Section 1.3, and for the BCM in Section 4.3 of this book (see, also, [104]). An analogous procedure is used here for problems in potential theory. The explicit equations for potential theory are available, for example, in Kane [68] (equations (17.25) and (17.34)). It should be noted here that the above mentioned equations in [68] were used to regularize the BIE and

HBIE, respectively, while use of these equations to obtain the primary variable and its derivative, at internal points close to the boundary of a body, is a new application of these equations. This approach, for the BNM for linear elasticity, is discussed later in Section 9.2.1.3 of this chapter.

9.1.2 HBNM: Coupling of HBIE with MLS Approximants

9.1.2.1 Coupled equations

This time, MLS approximations for the functions τ and $\partial u/\partial s$ (8.8, 8.20) are used in the HBIEs for 3-D potential theory (1.12, 1.13). The resulting HBNM equations are:

$$
\begin{aligned}
0 = & \sum_{i=1}^{N_c}\left\{\int_{\partial B_i}\frac{\partial G(\mathbf{x},\mathbf{y})}{\partial x_m}\left[\sum_{I=1}^{n_y}\Phi_I(\mathbf{y})\hat{\tau}_I - \sum_{I=1}^{n_x}\Phi_I(\mathbf{x})\hat{\tau}_I\right]dS(\mathbf{y})\right. \\
& - u_{,k}(\mathbf{x})\int_{\partial B_i}\frac{\partial G(\mathbf{x},\mathbf{y})}{\partial x_m}\left[n_k(\mathbf{y}) - n_k(\mathbf{x})\right]dS(\mathbf{y}) \\
& \left. - \int_{\partial B_i}\frac{\partial F(\mathbf{x},\mathbf{y})}{\partial x_m}\left[\sum_{I=1}^{n_y}\Phi_I(\mathbf{y})\hat{u}_I - \sum_{I=1}^{n_x}\Phi_I(\mathbf{x})\hat{u}_I - u_{,k}(\mathbf{x})(y_k - x_k)\right]dS(\mathbf{y})\right\}
\end{aligned}
$$

(9.12)

$$
\begin{aligned}
0 = & \sum_{i=1}^{N_c}\left\{\int_{\partial B_i}\frac{\partial G(\mathbf{x},\mathbf{y})}{\partial n(\mathbf{x})}\left[\sum_{I=1}^{n_y}\Phi_I(\mathbf{y})\hat{\tau}_I - \sum_{I=1}^{n_x}\Phi_I(\mathbf{x})\hat{\tau}_I\right]dS(\mathbf{y})\right. \\
& - u_{,k}(\mathbf{x})\int_{\partial B_i}\frac{\partial G(\mathbf{x},\mathbf{y})}{\partial n(\mathbf{x})}\left[n_k(\mathbf{y}) - n(\mathbf{x})\right]dS(\mathbf{y}) \\
& \left. - \int_{\partial B_i}\frac{\partial F(\mathbf{x},\mathbf{y})}{\partial n(\mathbf{x})}\left[\sum_{I=1}^{n_y}\Phi_I(\mathbf{y})\hat{u}_I - \sum_{I=1}^{n_x}\Phi_I(\mathbf{x})\hat{u}_I - u_{,k}(\mathbf{x})(y_k - x_k)\right]dS(\mathbf{y})\right\}
\end{aligned}
$$

(9.13)

respectively, where $\Phi_I(\mathbf{x})$ and $\Phi_I(\mathbf{y})$ are the contributions from the I_{th} node to the collocation point $(\mathbf{x})$ and field point $(\mathbf{y})$, respectively, with n_x and n_y nodes in their respective domains of dependence.

The discretized form of the potential gradient equation (1.14) at a source point $\mathbf{x}$ is:

$$
\nabla u(\mathbf{x}) = \mathbf{n}(\mathbf{x})\sum_{I=1}^{n_x}\Phi_I(\mathbf{x})\hat{\tau}_I + \mathbf{t}_1(\mathbf{x})\sum_{I=1}^{n_x}\Psi_I^{(1)}(\mathbf{x})\hat{u}_I + \mathbf{t}_2(\mathbf{x})\sum_{I=1}^{n_x}\Psi_I^{(2)}(\mathbf{x})\hat{u}_I
$$

(9.14)

The gradient of u from equation (9.14), in global coordinates, is used in equations (9.12) and (9.13).

9.1.2.2   Discretized equations

Equation (9.13) is used to solve boundary value problems in potential theory. As described above in Section 9.1.1.2, nonsingular integrals in (9.13) are evaluated by Gaussian quadrature while weakly singular integrals are evaluated by the procedure outlined in Section 9.1.1.2. The discretized version of equation (9.13) has the same form as (9.8) and boundary conditions are imposed in the manner described above in Section 9.1.1.3.

9.1.3   Numerical Results for Dirichlet Problems on a Sphere

The BNM [26] and the HBNM [27] are used to solve Dirichlet problems on a sphere. (Results for example problems on cubes are available in Chati and Mukherjee [26]). The exact solutions presented below have been used to evaluate the performance of the various parameters of the BNM and the HBNM. Dirichlet problems are posed with these solutions imposed (in turn) on the surface of a solid sphere, and the normal derivatives of the potential are computed on the sphere surface. Potential gradients at internal points are also computed in some cases. The complete sphere is modeled in all cases.

Exact solutions.

- Linear solution

$$u = x_1 + x_2 + x_3 \tag{9.15}$$

- Quadratic solution-one

$$u = x_1 x_2 + x_2 x_3 + x_3 x_1 \tag{9.16}$$

- Quadratic solution-two

$$u = -2(x_1)^2 + (x_2)^2 + (x_3)^2 \tag{9.17}$$

- Cubic solution

$$u = x_1^3 + x_2^3 + x_3^3 - 3x_1^2 x_2 - 3x_2^2 x_3 - 3x_3^2 x_1 \tag{9.18}$$

- Trigonometric solution

$$u = \frac{2r^2}{R^2} \cos^2 \phi - \frac{2r^2}{3R^2} - \frac{1}{3} \tag{9.19}$$

where R is the radius of the sphere, ϕ is the angle measured from the x_3 axis and $r^2 = x_1^2 + x_2^2 + x_3^2$.

Error measures. Several error measures have been employed in order to assess the accuracy of the numerical solutions. These are given below.

- Local

$$\varepsilon_{local} = \frac{\phi_i^{(num)} - \phi_i^{(exact)}}{\phi_i^{(exact)}} \times 100 \quad \% \qquad (9.20)$$

- Global One (node-based global percentage L_2 error - used for BNM results)

$$\varepsilon_{global} = \frac{1}{|\phi_i|_{max}} \sqrt{\frac{1}{N_B} \sum_{i=1}^{N_B} (\phi_i^{(num)} - \phi_i^{(exact)})^2} \quad \% \qquad (9.21)$$

- Global Two (integrated L_2 error - used for HBNM results)

$$\epsilon(\phi) = \frac{\int_A (\phi^{(num)} - \phi^{(exact)})^2 dA}{\int_A (\phi^{(exact)})^2 dA} \times 100 \quad \% \qquad (9.22)$$

In the above, ϕ is a generic function, ϕ_i is its nodal value at node i and $\phi^{(num)}$ is its numerical value.

Position of collocation nodes. One node per cell has been used for all the numerical examples presented in this section. It has been shown (from numerical experiments) in [26, 27] that, as expected, placement of the collocation node at the centroid of the triangle in the parent space yields excellent results. This nodal placement has been used in all the examples presented below.

9.1.3.1 Results from the BNM

A variety of problems have been solved on a sphere. The usual curvilinear coordinates θ and ϕ are used. As mentioned above, Dirichlet boundary conditions corresponding to the exact solution have been imposed on the surface of the sphere. Numerical results have been obtained using linear (T3) triangles and quadratic (T6) triangles for interpolating the geometry. To carry out the integration, each of these triangles are mapped into a unit triangle in the parent space (see Figure 9.2). The results have been obtained for four different meshes. The crude mesh contains 72 and the fine mesh 288 cells (over the surface of the sphere). Each of the meshes has two versions - using linear (T3) and quadratic (T6) triangles, respectively. These results are taken from [26].

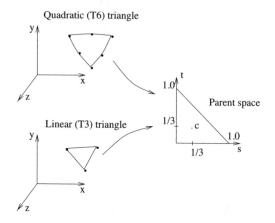

Figure 9.2: Mapping of linear (T3) and quadratic (T6) triangles onto the parent space (from [26])

Choice of compact support of weight function. The parameter d_I (the compact support of a weight function) needs to be chosen such that a 'reasonable' number of nodes lie in the domain of dependence of an evaluation point E. In Figure 9.3, the global one error (global percentage L_2 error from (9.21)) in τ has been recorded for increasing d_I. It is observed that choosing the smallest possible d_I yields the lowest L_2 error. For the optimum choice of d_I, there are about $12 - 14$ nodes in the range of influence of each node, which is about $2m - 3m$ for a quadratic polynomial basis $m = 6$.

Choice of weight function and polynomial basis. Various weight functions, proposed in the literature, are given in equations (8.22 -8.25). Tables 9.1, 9.2 and 9.3 present a convergence study that has been carried out for imposed linear, quadratic-one and cubic solutions, respectively (see (9.15 - 9.18)), for various weight functions proposed in the literature. It can be seen that a crude mesh with 72 quadratic (T6) cells already yields acceptable results. Also, the Gaussian weight function seems to have an edge over the other weight functions used. The results for the linear and quadratic-one solutions have been obtained using a polynomial basis $m = 6$, while, for the cubic solution, polynomial bases $m = 6$ and $m = 10$ have been used. It can be seen from Table 9.3 that using a higher order basis has only a marginal effect on the solution, except for the case with the fine mesh with T6 triangles. In fact, various researchers feel that it is best to use lower order polynomial bases in mesh-free methods.

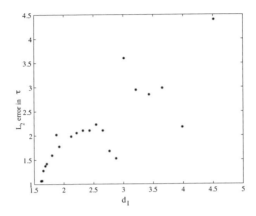

Figure 9.3: Effect of changing d_I on the global L_2 error in τ for a prescribed linear solution (from [26])

Weight Functions	Crude Mesh (T3)	Fine Mesh (T3)	Crude Mesh (T6)	Fine Mesh (T6)
WFA	4.237	2.083	1.067	0.497
WFB	4.237	2.086	1.069	0.539
WFC	5.029	2.104	1.031	0.584
WFD	4.057	2.253	1.246	0.757

Table 9.1: L_2 error in τ for the linear Dirichlet problem (from [26])

Weight Functions	Crude Mesh (T3)	Fine Mesh (T3)	Crude Mesh (T6)	Fine Mesh (T6)
WFA	4.022	1.788	1.696	0.886
WFB	4.022	1.789	1.697	1.011
WFC	5.314	1.785	1.722	1.199
WFD	5.391	1.893	2.103	1.942

Table 9.2: L_2 error in τ for the quadratic-one Dirichlet problem (from [26])

Weight	Crude Mesh (T3)		Fine Mesh (T3)	
Functions	$m = 6$	$m = 10$	$m = 6$	$m = 10$
WFA	5.641	5.944	2.284	2.319
WFB	5.641	5.944	2.285	2.318
WFC	6.007	5.525	2.298	2.301
WFD	7.896	5.674	2.345	2.318
Weight	Crude Mesh (T6)		Fine Mesh (T6)	
Functions	$m = 6$	$m = 10$	$m = 6$	$m = 10$
WFA	2.018	2.962	0.909	0.485
WFB	2.019	2.963	1.002	0.485
WFC	2.254	3.003	1.120	0.574
WFD	2.456	3.570	1.733	0.796

Table 9.3: L_2 error in τ for the cubic Dirichlet problem (from [26])

Coordinates	τ_{num}	τ_{exact}	Local Error
(2.0,0.0,0.0)	0.0	0.0	-
(1.5,0.866,1.0)	3.650	3.665	-0.413
(0.5,0.866,1.732)	2.852	2.799	1.897
(1.0,1.0,1.414)	3.847	3.828	0.475
(1.732,1.0,0.0)	1.689	1.732	-2.484
(1.414,1.414,0.0)	1.998	2.000	-0.118
(0.707,0.707,1.732)	3.053	2.949	3.513
(1.225,1.225,1.0)	3.912	3.949	-0.954

Table 9.4: Local error in τ at boundary points for the quadratic-one Dirichlet problem (from [26])

Results at boundary points other than at nodes. Upon solving a Dirichlet boundary value problem using the BNM, the value of τ is known at the N_B boundary nodes. However, to compute τ at any boundary location other than the collocation nodes, the MLS approximants, equation (8.8), need to be used. Table 9.4 presents the results for τ at a few points on the boundary that are not the collocation nodes. The results are obtained upon imposing a quadratic-1 solution on a crude mesh (72 T6 cells) with the Gaussian weight function. It can be seen that the accuracy of the numerical solution is well within acceptable limits.

Results at internal points. Figures 9.4 (a) and (b) show variation in the potential and its directional derivative at points inside the sphere. The Dirichlet boundary value problem is solved upon imposing the cubic solution on a crude mesh (72 T6 cells) with the Gaussian weight function and a quadratic basis ($m = 6$). The gradient is dotted with the diagonal ($x_1 = x_2 = x_3$) in order to get the directional derivative along this line. Values of u and ∇u, at internal points that are close to the surface of the body, are obtained with the usual BNM (without modification) and by a new application of equations (17.25) and (17.34) in Kane [68] (with modification). It is clear that this modification is essential for success of this method at computing derivatives of the potential function at internal points that are close to the boundary of a body. This issue, for the BEM for elasticity problems, is discussed in detail in Section 1.3 of this book, and for the BNM for elasticity in Section 9.2.1.3 later in this chapter.

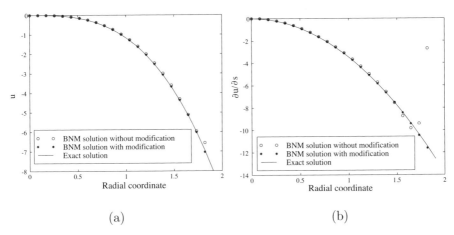

(a)   (b)

Figure 9.4: Variation of the potential u and tangential derivative of u, along the line $x_1 = x_2 = x_3$, with a prescribed cubic solution (from [26])

Comparison of BEM and BNM. The results obtained by the BNM have also been compared with those from the conventional BEM for a Dirichlet problem on a sphere with the exact trigonometric solution as given in equation (9.19). Table 9.5 compares the wall-clock times from a serial BEM code with a serial and a parallel version of the BNM code. The parallel version is an early one in which only the assembly phase of the BNM matrices, not the solution phase, has been parallelized. The parallel BNM code is run on 4, 16, and 32 processors using the message passing interface (MPI) standard on the IBM SP2 (R6000 architecture, 120 MHz P2SC Processor). Certainly, the serial BNM is considerably slower than the serial BEM and this is because the approximation functions in the BNM need to be generated at each point unlike in the BEM. (Similar observations have been made by other researchers regarding the performance of the EFG compared to the FEM). One possible remedy is an accelerated BNM (Kulkarni et al. [77]). Also, the BNM is very easy to parallelize,

Meshes	Serial BEM	Serial BNM	Parallel BNM		
			4 Procs	16 Procs	32 Procs
72 T6 cells	3.3 secs	29.5 secs	10.0 secs	2.4 secs	1.3 secs
128 T6 cells	9.3 secs	106.3 secs	35.5 secs	7.6 secs	4.4 secs
288 T6 cells	47.5 secs	690.7 secs	249.0 secs	53.0 secs	27.3 secs

Table 9.5: Comparison of wall-clock times for the BNM and BEM for a Dirichlet problem on a sphere (from [26])

and, as shown in Table 9.5, parallel versions drastically reduce wall-clock times. Figure 9.5 shows a comparison of the L_2 error for the BEM and the BNM as a function of the number of boundary nodes. The L_2 error in τ is defined here as:

$$e(\tau) = \frac{\sum_{i=1}^{N_B}(\tau_i^{(num)} - \tau_i^{(exact)})^2}{\sum_{i=1}^{N_B}(\tau_i^{(exact)})^2} \tag{9.23}$$

It can be clearly seen that the two methods yield comparable results and have almost identical rates of convergence.

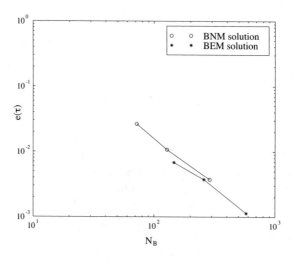

Figure 9.5: Convergence of the BNM and the BEM with a prescribed trigonometric solution (from [26])

9.1.3.2   Results from the HBNM

This section follows the layout of Section 9.1.3.1 above. Once again, Dirichlet problems have been solved with various forms of the potential function (from

Collocation point : P $(x_{1P}, x_{2P}, x_{3P}) \equiv (R, \theta_P, \phi_P)$	
Field point : Q $(x_{1Q}, x_{2Q}, x_{3Q}) \equiv (R, \theta_Q, \phi_Q)$	
Curvilinear coordinates between P and Q :	
$\tilde{s}_1 = R(\phi_Q - \phi_P)$; $\tilde{s}_2 = R(\theta_Q - \theta_P)$	
Exact geodesic	Approximate geodesic
Ψ : Angle between P & Q	$d = \sqrt{(\tilde{s}_1^2 + \tilde{s}_2^2)}$
$\Psi = \arccos(\bar{\mathbf{r}}_P \cdot \bar{\mathbf{r}}_Q / R^2)$	
$d = R\Psi$	

Table 9.6: Exact versus approximate geodesics on the surface of a sphere (from [27])

equations (9.15 - 9.19)) prescribed on the surface of a sphere. The Gaussian weight function (8.22) is used in all cases. Also, for all the numerical examples presented in this section, 72 T6 triangular cells have been used on the surface of a sphere, with one node per cell placed at the centroid of each triangle in the parent space (Figure 9.2). Overall, this arrangement yields the best numerical results. These results are taken from [27].

Geodesics. In order to construct the interpolating functions using MLS approximants, it is necessary to compute the geodesic on the bounding surface of a body. Computation of geodesics can get quite cumbersome on a general curved surface described by splines. For the sphere, the exact geodesic between two points **x** (collocation node) and **y** (field point) is the length of the arc between these points on the great circle containing them. However, a very simple approximation to the geodesic would be to use the "Euclidean" distance between points **x** and **y**. Table 9.6 summarizes the procedure for computing the exact and approximate geodesic on a sphere, while Table 9.7 presents a comparison in the (global two) L_2 errors for linear, quadratic, cubic and trigonometric solutions imposed on the sphere. For this specific example, with idealized geometry and boundary conditions, it can be seen that the errors remain reasonably small even if the approximate geodesic is used to replace the exact one. Thus, it is expected that the computation of geodesics on complicated shapes will not be a hindrance towards using the present methodology.

Range of influence of nodes. Another important feature of the MLS approximants is the range of influence associated with each node. The parameter which controls the so-called "compact support" associated with each node is d_I. In this work the parameter d_I is chosen to be *nonhomogeneous* in the sense that each evaluation point has an *identical* number of nodes in its domain of dependence. Now, for a given polynomial basis (e.g. Linear/Quadratic/Cubic),

Exact solution	Exact geodesic	Approximate geodesic
Linear	0.0717 %	0.109 %
Quadratic	0.313 %	0.715 %
Cubic	1.665 %	4.059 %
Trigonometric	0.172 %	0.279 %

Table 9.7: Global two L_2 error in τ for Dirichlet problems on a sphere with the exact and an approximate computation of geodesics (from [27])

the number of nodes n in the domain of dependence of each evaluation point becomes the parameter of interest.

The parameters d_I are chosen as follows. For a given n, let $\mathcal{S}$ be the set of nodes in the domain of dependence of a particular evaluation point E. The values of d_I for all nodes in $\mathcal{S}$ are set equal to d_{max}, where d_{max} is the distance from E, along the geodesic, of the node in $\mathcal{S}$ which is farthest from E.

Figures 9.6 (a) and (b) show the effect of varying the number of nodes n for a linear basis ($m = 3$) and a quadratic basis ($m = 6$), respectively, for the cubic and trigonometric solutions. It is observed that the lowest value of L_2 errors is obtained for $n \in (2m, 3m)$. This fact has also been observed for the BNM.

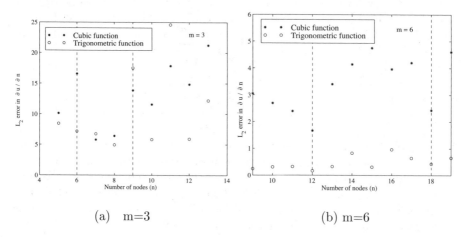

(a)   m=3   (b) m=6

Figure 9.6: Global two L_2 error in τ for varying number of points in the domain of dependence of an evaluation point E: (a) linear polynomial basis ($m = 3$); (b) quadratic polynomial basis ($m = 6$) (72 T6 cells with one node per cell) (from [27])

Results at internal points. Figures 9.7 (a) and (b) show variation in the potential and its x_1 derivative, respectively, for points along the x_1 axis inside the sphere. The Dirichlet boundary value problem is solved upon imposing the

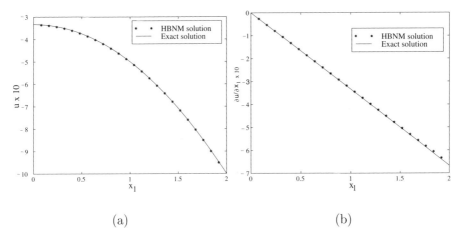

(a)   (b)

Figure 9.7: Variation of (a) potential and (b) $\partial u/\partial x_1$ along the x_1 axis for a sphere with the prescribed trigonometric solution (from [27])

trigonometric solution on a cell configuration consisting of 72 T6 cells with one node per cell and a quadratic basis ($m = 6$). It is seen from these figures that the HBNM solutions match the exact solutions within plotting accuracy for both u and $\partial u/\partial x_1$. Once again, the reader is reminded that a technique for dealing with nearly singular integrals in linear elasticity is described in detail in Section 1.3 (for the BEM), and in Section 9.2.1.3 (for the BNM) later in this chapter. An analogous method for potential theory has been employed in order to get accurate results at internal points that are close to the surface of the sphere in Figure 9.7.

Comparison of BEM and HBNM. The results obtained by the HBNM have also been compared with those from the conventional BEM for a Dirichlet problem on a sphere with the trigonometric solution (equation (9.19)). Figure 9.8 presents a comparison in the L_2 error in τ (as defined in (9.23)) for the HBNM and BEM, as functions of the (global) number of nodes. This figure shows that the two methods yield comparable results and have similar rates of convergence. The HBNM solution, however, is more accurate than the BEM solution for this example.

9.2 Linear Elasticity in Three Dimensions

9.2.1 BNM: Coupling of BIE with MLS Approximants

9.2.1.1 Coupled equations

As before for the case of potential theory (Section 9.1.1.1), the bounding surface ∂B of a body B is partitioned into N_c cells ∂B_k and MLS approximations for

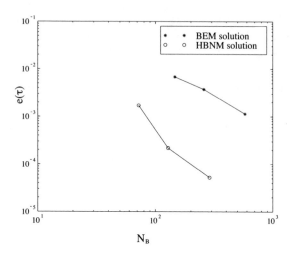

Figure 9.8: Comparison of $e(\tau)$ (L_2 error in $\partial u/\partial n$) for the HBNM and the conventional BEM for a Dirichlet problem on a sphere (from [27])

the functions u_i and τ_i (8.8) are used in the usual BIE for 3-D linear elasticity (1.25). The result is:

$$
0 = \sum_{m=1}^{N_c} \int_{\partial B_m} \left[U_{ik}(\mathbf{x}, \mathbf{y}) \sum_{I=1}^{n_y} \Phi_I(\mathbf{y}) \hat{\tau}_{iI} \right.
$$
$$
\left. - T_{ik}(\mathbf{x}, \mathbf{y}) \left\{ \sum_{I=1}^{n_y} \Phi_I(\mathbf{y}) \hat{u}_{iI} - \sum_{I=1}^{n_x} \Phi_I(\mathbf{x}) \hat{u}_{iI} \right\} \right] dS(\mathbf{y}) \qquad (9.24)
$$

where $\Phi_I(\mathbf{x})$ and $\Phi_I(\mathbf{y})$ are the contributions from the I_{th} node to the collocation point $\mathbf{x}$ and field point $\mathbf{y}$ respectively. Also, as before, n_y nodes lie in the domain of dependence of the field point $\mathbf{y}$ and n_x nodes lie in the domain of dependence of the source point $\mathbf{x}$. When $\mathbf{x}$ and $\mathbf{y}$ belong to the same cell, the cell is treated as a singular cell and the special techniques described in Section 9.1.1.2 are used to carry out the integrations. Otherwise, regular Gaussian integration is used.

9.2.1.2   Discretized equations and boundary conditions

This discussion closely follows that on potential theory in Section 9.1.1.3. The discretized form of (9.24) has the same form as (9.8), but this time with $3N_B$ equations for the $6N_B$ quantities $\hat{u}_i, \hat{\tau}_i$, $i = 1, 2, 3$, at N_B boundary nodes. The other $6N_B$ quantities of interest are u_i, τ_i, $i = 1, 2, 3$, at N_B boundary nodes; $3N_B$ of which are prescribed by the boundary conditions. The remaining $6N_B$ equations, that relate the nodal approximations of u_i and τ_i to their nodal values, have the same form as (8.11). The complete system of equations is solved in a manner that is very similar to that described in Section 9.1.1.3 for potential theory.

9.2.1.3 Displacements and stresses at internal points

The displacement at a point $\boldsymbol{\xi} \in B$ is obtained from (1.17) or (1.21), the displacement gradient from (1.27) or (1.28), and the stress (directly) from (1.30). In each case, of course, (8.8) must be used.

As mentioned several times before (see Sections 1.3, 4.3), *evaluation of displacements and stresses at internal points, that are very close to the bounding surface of a body, requires special care.* For this purpose, the BNM versions of continuous equations such as (1.50) and (1.52) are needed. These are [104]:

$$
\begin{aligned}
u_k(\boldsymbol{\xi}) \;=\;& u_k(\hat{\mathbf{x}}) \\
+& \sum_{m=1}^{N_c} \int_{\partial B_m} \left[U_{ik}(\boldsymbol{\xi}, \mathbf{y}) \sum_{I=1}^{n_y} \Phi_I(\mathbf{y}) \hat{\tau}_{iI} - T_{ik}(\boldsymbol{\xi}, \mathbf{y}) \left\{ \sum_{I=1}^{n_y} \Phi_I(\mathbf{y}) \hat{u}_{iI} - u_i(\hat{\mathbf{x}}) \right\} \right] dS(\mathbf{y})
\end{aligned}
$$

$$
(9.25)
$$

$$
\begin{aligned}
\sigma_{ij}(\boldsymbol{\xi}) \;=\;& \sigma_{ij}(\hat{\mathbf{x}}) + \int_{\partial B} D_{ijk}(\boldsymbol{\xi}, \mathbf{y}) \left[\sum_{I=1}^{n_y} \Phi_I(\mathbf{y}) \hat{\tau}_{kI} - \sigma_{km}(\hat{\mathbf{x}}) n_m(\mathbf{y}) \right] dS(\mathbf{y}) \\
& - \int_{\partial B} S_{ijk}(\boldsymbol{\xi}, \mathbf{y}) \left[\sum_{I=1}^{n_y} \Phi_I(\mathbf{y}) \hat{u}_{kI} - u_k(\hat{\mathbf{x}}) - u_{k,\ell}(\hat{\mathbf{x}})(y_\ell - \hat{x}_\ell) \right] dS(\mathbf{y})
\end{aligned}
$$

$$
(9.26)
$$

Equation (9.26)) requires the displacement gradients and stresses at the target point $\hat{\mathbf{x}}$ (in addition to the usual displacements and tractions at field points $\mathbf{y}$). Displacement gradients on the surface of the body are obtained as part of the BNM solution of the original boundary value problem. This procedure, adapted from Lutz et al. [89] for the BEM, is described in detail in Section 1.2.2.1 of this book (see, also, [27]).

9.2.2 HBNM: Coupling of HBIE with MLS Approximants

9.2.2.1 Coupled equations

The starting point here are the HBIEs (1.40) and (1.41). The HBNM equations are:

$$
0 \;=\; \sum_{l=1}^{N_c} \int_{\partial B_l} D_{ijk}(\mathbf{x}, \mathbf{y}) \left[\sum_{I=1}^{n_y} \Phi_I(\mathbf{y}) \hat{\tau}_{kI} - \sum_{I=1}^{n_x} \Phi_I(\mathbf{x}) \hat{\tau}_{kI} \right] dS(\mathbf{y})
$$

$$- \quad \sigma_{km}(\mathbf{x}) \int_{\partial B_l} D_{ijk}(\mathbf{x}, \mathbf{y})(n_m(\mathbf{y}) - n_m(\mathbf{x}))\, dS(\mathbf{y})$$

$$- \int_{\partial B_l} S_{ijk}(\mathbf{x}, \mathbf{y}) \left[\sum_{I=1}^{n_y} \Phi_I(\mathbf{y})\hat{u}_{kI} - \sum_{I=1}^{n_x} \Phi_I(\mathbf{x})\hat{u}_{kI} - u_{k,m}(\mathbf{x})(y_m - x_m) \right] dS(\mathbf{y})$$

$$(9.27)$$

and

$$0 \quad = \quad \sum_{l=1}^{N_c} \int_{\partial B_l} D_{ijk}(\mathbf{x}, \mathbf{y})n_j(\mathbf{x}) \left[\sum_{I=1}^{n_y} \Phi_I(\mathbf{y})\hat{\tau}_{kI} - \sum_{I=1}^{n_x} \Phi_I(\mathbf{x})\hat{\tau}_{kI} \right] dS(\mathbf{y})$$

$$- \quad \sigma_{km}(\mathbf{x}) \int_{\partial B_l} D_{ijk}(\mathbf{x}, \mathbf{y})n_j(\mathbf{x})(n_m(\mathbf{y}) - n_m(\mathbf{x}))\, dS(\mathbf{y})$$

$$- \int_{\partial B_l} S_{ijk}(\mathbf{x}, \mathbf{y})n_j(\mathbf{x}) \left[\sum_{I=1}^{n_y} \Phi_I(\mathbf{y})\hat{u}_{kI} - \sum_{I=1}^{n_x} \Phi_I(\mathbf{x})\hat{u}_{kI} - u_{k,m}(\mathbf{x})(y_m - x_m) \right] dS(\mathbf{y})$$

$$(9.28)$$

Note that, as expected, taking the limit $\hat{\mathbf{x}} \in \partial B \to \mathbf{x} \in \partial B$ in the continuous equation (9.26) yields a corresponding regularized HBNM. The resulting equation is an alternate version of (9.27) (see (1.39) and (1.40)).

9.2.2.2   Discretization

Equations (9.27) and (9.28) are the HBNM equations for linear elasticity.

The procedure followed for discretization of (9.28) is quite analogous to the BNM case described before in Section 9.2.1.2. These equations are fully regularized and only contain nonsingular or weakly singular integrands. Nonsingular integrals are evaluated using the usual Gauss quadrature over surface cells, while the weakly singular integrals are evaluated using the procedure outlined in Section 9.1.1.2). Displacement gradients at a boundary point $\mathbf{x}$ are obtained by applying the procedure outlined in Section 1.2.2.1 and the stress components at this point are then obtained from Hooke's law. The discretized version of equation (9.28) has the generic form shown in (9.8).

9.2.3   Numerical Results

Numerical results from the standard and the new BNM, as well as from the HBNM, are presented in this section. The standard BNM results are obtained from (the BNM versions of) equations (1.17) and (1.30) while the new BNM results are obtained from equations (9.25) and (9.26). Three illustrative examples are considered in this section : a hollow sphere under internal pressure, a bimaterial sphere and a cube with a spherical cavity under tension (the 3-D Kirsch problem). These numerical results are taken from [25], [104] and [27].

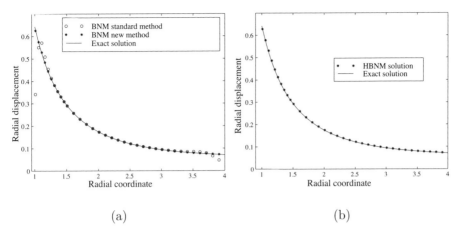

(a)   (b)

Figure 9.9: Radial displacement as a function of the radial coordinate in a hollow sphere: (a) from the standard and the new BNM (b) from the HBNM, together with the exact solution (from [104] and [27])

9.2.3.1 Hollow sphere under internal pressure

The inner and outer radii of the sphere are 1 and 4 units, respectively, the internal pressure is 1, $E = 1$ and $\nu = 0.25$. 72 quadratic $T6$ triangular cells (with one node at the centroid of each cell) are used on each surface of the hollow sphere. The entire surface of the sphere is modeled here, with tractions prescribed over both the inner and outer surfaces of the sphere. The resulting singular matrices are regularized using the procedure described in [25] (see, also, [90]).

Internal displacements.   The radial displacement, as a function of the radial coordinate, from the standard as well as the new BNM, are shown in Figure 9.9 (a) while corresponding results from the HBNM appear in Figure 9.9 (b). The exact solution from [167] is also included in these figures. It is observed that the results from the standard BNM and HBNM are excellent as long as an internal point is not very close to one of the surfaces of the sphere; while those from the new BNM are very accurate everywhere, including at internal points that are close to the bounding surfaces of the sphere.

Internal stresses.   The radial and circumferential stresses from the new BNM and from the HBNM, together with the exact solutions from [167], appear in Figures 9.10 (a) and (b), respectively. The agreement between the exact and numerical solutions are observed to be excellent.

Finally, the stress solutions from the standard BNM are shown in Figure 9.11, together with details near the inner surface of the sphere in Table 9.8. It is observed that the stress solutions from the standard BNM are meaningless at

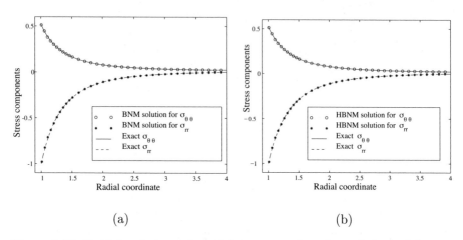

(a)   (b)

Figure 9.10: Radial and circumferential stresses as functions of the radial coordinate in a hollow sphere: (a) from the new BNM (b) from the HBNM, together with the exact solutions (from [104] and [27])

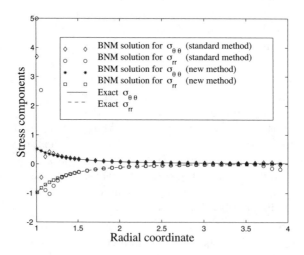

Figure 9.11: Radial and circumferential stresses as functions of the radial coordinate in a hollow sphere, from the new and standard BNM, together with exact solutions (from [104])

	σ_{rr}			$\sigma_{\theta\theta}$		
r	standard	new	exact	standard	new	exact
1.01	4.99057	-0.98020	-0.97012	3.68714	0.51168	0.50887
1.03	5.43126	-0.92753	-0.91379	1.08215	0.49365	0.48071
1.05	3.76287	-0.85789	-0.86168	-0.35304	0.46201	0.45465
1.07	1.39290	-0.79589	-0.81338	-0.36737	0.42834	0.43050
1.09	-0.18719	-0.74915	-0.76857	-0.01546	0.40084	0.40809
1.11	-0.90056	-0.71229	-0.72693	0.24717	0.37928	0.38727

Table 9.8: Radial and circumferential stresses as functions of the radial coordinate in a hollow sphere, from the new and standard BNM, together with the exact solutions, at points *very close* to the inner surface (from [104])

internal points very near the inner surface of the sphere, and an algorithm for improving these results, such as that presented in Section 9.2.1.3, is absolutely essential in this case.

9.2.3.2 Bimaterial sphere

The BNM has been extended to solve problems involving material discontinuities. Figure 9.12 (a) shows a schematic of two perfectly bonded spheres of two different materials. Numerical results for this model have been obtained by prescribing displacements on the outer boundary. One could also prescribe tractions over the entire outer surface and then appropriately modify the scheme presented in [25] for solving traction prescribed problems. This is planned for the future.

Upon prescribing a radial displacement u_0 on the outer surface of the sphere, the exact solution for the radial displacement and stresses in each material is given below. For material 1:

$$u_r^{(1)} = A_1 r \frac{1 - 2\nu_1}{E_1} \tag{9.29}$$

$$\sigma_{rr}^{(1)} = \sigma_{tt}^{(1)} = A_1 \tag{9.30}$$

where the constant A_1 is defined below. For material 2:

$$u_r^{(2)} = \frac{r}{E_2} \left[A_2(1 - 2\nu_2) - \frac{B_2}{2r^3}(1 + \nu_2) \right] \tag{9.31}$$

$$\sigma_{rr}^{(2)} = A_2 + \frac{B_2}{r^3} \quad ; \quad \sigma_{tt}^{(2)} = A_2 - \frac{B_2}{2r^3} \tag{9.32}$$

The constants are:

$$A_1 = A_2 + \frac{B_2}{R_1^3} \tag{9.33}$$

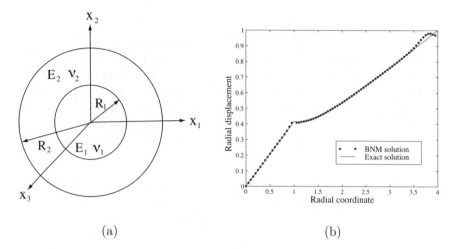

(a)   (b)

Figure 9.12: Bimaterial sphere subjected to uniform external radial displacement : (a) configuration (b) radial displacement along the x_1 axis from the new BNM together with the exact solution (from [25])

$$\alpha = \frac{E_1(1 - 2\nu_2)}{E_2(1 - 2\nu_1)} \quad ; \quad \beta = \frac{E_1(1 + \nu_2)}{2E_2(1 - 2\nu_1)} \tag{9.34}$$

$$A_2 = \frac{u_0 E_2}{R_2 R_1^3} \frac{1 + \beta}{C(1 + \alpha)} \quad ; \quad B_2 = -\frac{u_0 E_2}{R_2 C} \tag{9.35}$$

$$C = \frac{(1 - 2\nu_2)(1 + \beta)}{R_1^3(1 - \alpha)} + \frac{1 + \nu_2}{2R_2^3} \tag{9.36}$$

The material and geometric parameters chosen for the two materials are :

- for material 1, $E_1 = 1.0, \nu_1 = 0.28, R_1 = 1.0$

- for material 2, $E_2 = 2.0, \nu_2 = 0.33, R_2 = 4.0$.

A constant radial displacement is prescribed on the outer boundary of material 2 ($u_0 = 1.0$). Figure 9.12 (b) shows a comparison of the numerical solution and the exact solution for the radial displacement within the two materials. Figures 9.13 (a) and (b) show the radial and circumferential stress, respectively, along the line x_1 axis. It can be seen that the jump in the circumferential stress at the bimaterial interface is very well captured by the new BNM.

9.2.3.3 3-D Kirsch problem

The 3-D Kirsch's problem consists of a cube with a small spherical cavity subjected to far field uniform tension (Figure 9.14 (a)). The material and geometric parameters are : $E = 1.0, \nu = 0.25$, cutout radius $a = 1.0$, side of cube $2b = 20.0$. Again, the loading is applied without restraining any rigid body

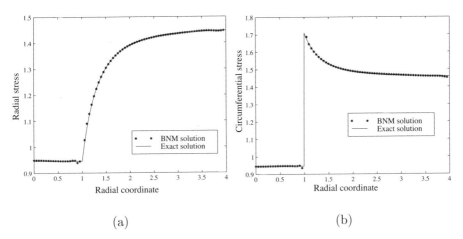

(a)   (b)

Figure 9.13: Bimaterial sphere subjected to uniform external radial displacement : (a) radial stress σ_{rr} (b) circumferential stress $\sigma_{\theta\theta}$ along the x_1 axis from the new BNM, together with the exact solutions (from [25])

modes and the scheme presented in [25] is used to obtain meaningful numerical results.

The exact solution for the normal stress σ_{33}, for points in the plane $x_3 = 0$, is given as [167]:

$$\sigma_{33} = \sigma_0 \left[1 + \frac{4 - 5\nu}{2(7 - 5\nu)} \left(\frac{a}{r} \right)^3 + \frac{9}{2(7 - 5\nu)} \left(\frac{a}{r} \right)^5 \right] \qquad (9.37)$$

Figure 9.14 (b) shows a comparison between the new BNM, HBNM and the exact solution for the (normalized) normal stress σ_{33}/σ_0 along the x_1-axis. Again, it can be clearly seen that the BNM and HBNM solutions are in excellent agreement with the analytical solution. The cell structure consists of 96 Q4 cells modeling the cube and 72 T6 cells modeling the spherical cavity, again with one node per cell. It should be noted that the algorithm presented in Section 9.2.1.3 is essential for obtaining accurate values of stresses near the surface of the cavity.

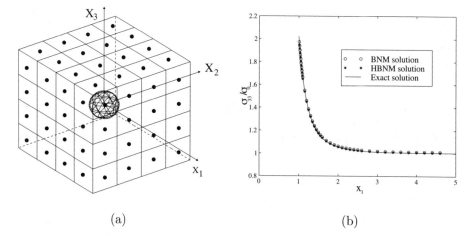

(a)   (b)

Figure 9.14: 3-D Kirsch problem : (a) configuration (b) σ_{33}/σ_0 along the x_1 axis from the new BNM and the HBNM, together with the exact solution (from [25, 27])

Chapter 10

ADAPTIVITY FOR 3-D POTENTIAL THEORY

The subject of this chapter is error analysis and adaptivity with the BNM, as applied to problems in 3-D potential theory. The idea of using hypersingular residuals, to obtain local error estimates for the BIE, was first proposed by Paulino [122] and Paulino et al. [123]. This idea has been applied to the collocation BEM (Paulino et al. [123], Menon et al. [96] and Paulino et al. [127]); and has been discussed in detail in Chapter 2 of this book. This idea, applied to the BCM, has appeared in Mukherjee and Mukherjee [111], and is presented in Chapter 7 of this book. This idea has also been applied to the BNM, for problems in 3-D potential theory and linear elasticity [28]. Applications in 3-D potential theory is the subject of this chapter, while applications in 3-D elasticity are discussed in the next chapter - the last one in this book.

10.1 Hypersingular and Singular Residuals

10.1.1 The Hypersingular Residual

Let the BNM (equation (9.1)) for potential theory be written in operator form as:

$$\mathcal{L}_{BNM}(u, \tau) = 0 \qquad (10.1)$$

and its *numerical solution* be (u^*, τ^*). Also, the HBNM (equation (9.12)) is written in operator form as:

$$\mathcal{L}_{HBNM}(u, \tau) = 0 \qquad (10.2)$$

The hypersingular residual in the potential gradient $u_{,j}$ is defined as,

$$r_j \equiv residual(u_{,j}) = \mathcal{L}_{HBNM}(u^*, \tau^*) \qquad (10.3)$$

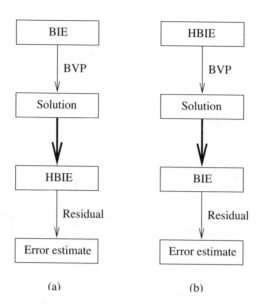

(a)   (b)

Figure 10.1: Interchange of BIE and HBIE (a) hypersingular residual (b) singular residual (from [28])

and is calculated from equation (9.12).

This idea is illustrated in Figure 10.1(a)

It has been proved in [96] and [127] for the BIE that, under certain favorable conditions, real positive constants c_1 and c_2 exist such that:

$$c_1 r \le \epsilon \le c_2 r \qquad (10.4)$$

where r is some scalar measure of a hypersingular residual and ϵ is a scalar measure of the exact local error. Thus, a hypersingular residual is expected to provide a good estimate of the local error on a boundary element. It should be mentioned here that the definitions of the residuals used in [96] and [127] are analogous to, but different in detail from, the ones proposed in this chapter.

10.1.2   The Singular Residual

The argument for using the residuals as error estimates is symmetric (see Paulino [122], Paulino et al. [123]). Therefore, one can reverse the above procedure to define singular residuals by first solving the HBIE and then iterating with the BIE.

In this case, for potential theory, one gets from equation (9.13):

$$\mathcal{L}_{HBNM}(u^o, \tau^o) = 0 \qquad (10.5)$$

and from equation (9.1):

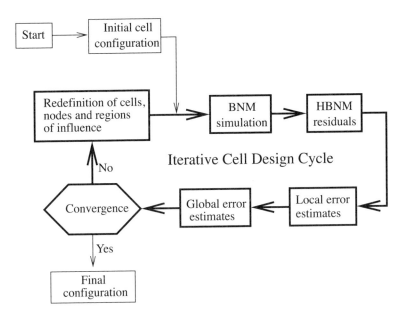

Figure 10.2: Typical self-adaptive iterative BNM algorithm (h−version) according to the scheme of Figure 10.1(a). The BNM equation used for solving the BVP is (9.1), and the HBNM equation used for residual computation is (9.12) (from ([28])

$$r \equiv residual(u) = |\mathcal{L}_{BNM}(u^o, \tau^o)| \qquad (10.6)$$

This idea is illustrated in Figure 10.1(b).

The above formulation for singular and hypersingular residuals is a generalization of the earlier work by Menon et al. [96] in the sense that Dirichlet, Neumann and mixed problems require separate prescriptions in [96], while the current work presents a unified residual formulation.

10.2 Error Estimation and Adaptive Strategy

There are similarities between adaptive techniques (e.g. h-version) for mesh-based methods (see Paulino et al. [126, 124]) and meshless methods. However, the latter set of methods provides substantially more flexibility in the (re-)discretization process than the former ones.

The h−version iterative self-adaptive procedure employed in this work is presented in the flowchart - Figure 10.2. The goal is to efficiently develop a final cell configuration which leads to a reliable numerical solution, in as simple a manner as possible.

10.2.1   Local Residuals and Errors - Hypersingular Residual Approach

From equation (10.3):

$$r_j = residual(u_{,j}) \tag{10.7}$$

A scalar residual measure is defined as:

$$r = r_j r_j \tag{10.8}$$

The exact local error in the gradient, $u_{,j}$, is defined as:

$$\epsilon_j = u_{,j}^{(exact)} - u_{,j}^{(numerical)} \tag{10.9}$$

and the corresponding scalar measure is defined as:

$$\epsilon = \epsilon_j \epsilon_j \tag{10.10}$$

Equations (10.8) and (10.10) are used to calculate the hypersingular residual and exact error, respectively, in the gradient $u_{,j}$, at each node, for problems in potential theory.

10.2.2   Local Residuals and Errors - Singular Residual Approach

The singular residual is defined in an analogous fashion. From equation (10.6):

$$r = residual(u) \tag{10.11}$$

and the exact local error in u is defined as,

$$e = |u^{(exact)} - u^{(numerical)}| \tag{10.12}$$

Here, r and ϵ are themselves scalar measures of the residual and exact error, respectively. *Equations (10.6) and (10.12) are used to obtain the singular residual and exact error, respectively, in the potential u, at each node, for problems in potential theory.* These equations are presented here for the sake of completeness.

The local error measure (equation (10.12)) is also used for $\partial u/\partial n$ at points on the surface of a cube (see examples of Section 10.3). This quantity is defined as:

$$e\left(\frac{\partial u}{\partial n}\right) = \left| \frac{\partial u}{\partial n}^{(exact)} - \frac{\partial u}{\partial n}^{(numerical)} \right| \tag{10.13}$$

This error measure is only used in Figure 10.8.

The scalar residual measures, defined above, evaluated at nodes, are used as error estimators. In all the adaptivity examples presented in this chapter, one node is used for each cell and is placed at its centroid. The scalar residual measure at this centroidal node is used as an error estimator for that cell. A comparison of the residual r and exact error ϵ demonstrates the effectiveness of residuals as error estimates.

10.2.3   Cell Refinement Criterion

A simple criterion for cell refinement consists of subdividing the cells for which the error indicator is larger than a certain reference value. In this work, the reference quantity is taken as the average value of the error indicator (here the average residual) given by:

$$\bar{r} = \frac{1}{N_n} \sum_{i=1}^{N_n} r^{(i)} \tag{10.14}$$

where N_n is the total number of nodes. If the inequality:

$$r > \gamma \, \bar{r} \tag{10.15}$$

is satisfied, then the cell is subdivided into four pieces (see Figure 10.3). The parameter γ in equation (10.15) is a weighting coefficient that controls the "cell refinement velocity." The standard procedure consists of using $\gamma = 1.0$. If $\gamma > 1.0$, then the number of cells to be refined is less than with $\gamma = 1.0$. According to Figure 10.2, the numerical solution of the next iterative step is expected to be more accurate than that of the current step; however, the increase on the total number of cells is comparatively small when $\gamma > 1.0$.

If $\gamma < 1.0$, then the number of cells to be refined is larger than that with $\gamma = 1.0$. The advantage in this case is that the refinement rate increases, however, the computational efficiency may decrease owing to likely generation of an excessive number of cells. *An alternative procedure, for a ONE-step refinement, is presented in Section 10.4 of this chapter.*

10.2.4   Global Error Estimation and Stopping Criterion

Global L_2 error.   A global L_2 error, on a panel, or over the whole boundary ∂B, is defined as

$$\bar{\epsilon}(\phi) = \frac{\int_A (\phi^{(exact)} - \phi^{(numerical)})^2 \, dA}{\int_A (\phi^{(exact)})^2 \, dA} 100\% \tag{10.16}$$

where ϕ is a variable of interest and A is the area of a panel or of the whole surface ∂B. These global errors are used in many of the tables that are presented later in this paper.

An indication of overall convergence may be obtained by evaluating either $\bar{r}$ (equation (10.14)) or $\bar{\epsilon}$ from equation (10.16). Of course, equation (10.16) is only useful for test examples in which the exact solution is known.

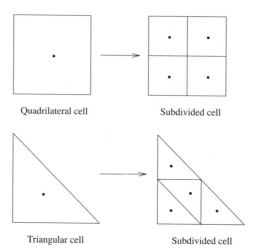

Figure 10.3: Cell refinement for quadrilateral and triangular cells with one node per cell (from ([28])

Stopping criterion.   For generic problems where the exact solution is not available (e.g. most engineering problems), cell refinement (see Figure 10.3) can be stopped when:

$$\bar{r} \leq r_{global} \tag{10.17}$$

where r_{global} has a preset value, which depends on the overall level of accuracy desired. The goal of the adaptive procedure is to obtain well-distributed (i.e. near optimal) cell configurations. Ideally, as the iterative cell refinement progresses, the error estimates should decrease both locally and globally.

10.3   Progressively Adaptive Solutions: Cube Problem

The adaptive process illustrated by Figure 10.2 is applied to the representative example of a Dirichlet problem on a cube. Laplace's equation is solved using the BNM, and the (hypersingular) residuals are obtained using the HBNM, according to Figure 10.1(a).

This example, together with an elasticity example discussed in Chapter 11, permits assessment of various parameters of the adaptive strategy for meshless methods based on BIE techniques. Several aspects are investigated such as the quality of the adaptive solution obtained for scalar (potential theory) and vector field (elasticity theory) problems, performance of the method on problems with either pure or mixed boundary conditions, evaluation of the quality of error estimates obtained by means of hypersingular or singular residuals, sensitivity of the "final" solution with respect to the starting cell configuration (initial condition of the self-adaptive problem), and convergence properties.

10.3.1 Exact Solution

The following exact solution, which satisfies the 3-D Laplace's equation, is used in this example :

$$u = \sinh\left(\frac{\pi x_1}{2}\right) \sin\left(\frac{\pi x_2}{2\sqrt{2}}\right) \sin\left(\frac{\pi x_3}{2\sqrt{2}}\right) \tag{10.18}$$

Note that the solution (10.18) is symmetric with respect to y and z but that its dependence on x is different from its dependence on y or z. The appropriate value of u is prescribed on ∂B (Dirichlet problem) and $\partial u/\partial n$ is computed on ∂B. Because the exact solution cannot be represented in terms of polynomials, this is a proper test of the meshless method and the adaptivity procedure. A quadratic basis is used for the construction of the MLS interpolating functions, i.e. $m = 6$ (see equation (8.2)). The idea behind the adaptive procedure is to start with a rather crude cell configuration and carry out cell refinement in the region where the residual is large according to a certain criterion. Hence, the adaptivity results in this section have been obtained starting with two different relatively coarse initial cell configurations. This comparative procedure tests the sensitivity of the adaptive scheme with respect to the initial conditions.

10.3.2 Initial Cell Configuration # 1 (54 Surface Cells)

Figure 10.4(a) shows a discretization consisting of 54 rectangular cells with one (centroidal) node per cell. The boundary value problem is solved using the BNM (equation (9.1)). Then the results are used in the HBNM (equation (9.12)) to obtain the hypersingular residual. Figure 10.5 shows a comparison between the hypersingular residual (from equations (10.3) and (10.8)) and the exact local error ϵ in $u_{,j}$ (from equations (10.9) and (10.10)) computed for the initial configuration # 1 (Figure 10.4(a)) at each node on the surface. It can be clearly seen that the hypersingular residual tracks the exact error perfectly.

Cell refinement is carried out using $\gamma = 1.0$ in equation (10.15), and the resulting refined cell configuration consisting of 126 cells is shown in Figure 10.4(b). It can be seen from Figure 10.4(b) that the cell refinement occurs *only* at the corners where the exact error is the largest. This is an indication that the procedure for error estimation and adaptivity is moving in the right direction. Now, the boundary value problem is solved again using the BNM. Table 10.1 summarizes the various output parameters of the adaptivity procedure. It can be seen from Table 10.1 that excellent numerical results are obtained in a single step of the adaptivity process and hence the adaptive procedure is not continued further.

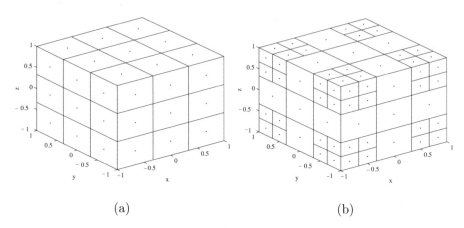

(a)   (b)

Figure 10.4: Cell configurations on the surface of a cube. (a) Initial configuration # 1: 54 surface cells. (b) First adapted step : 126 cells, obtained with $\gamma = 1$ (from ([28])

Output parameters	Initial	Final
Number of cells	54	126
$x = \pm 1$	1.4209 %	0.0238 %
$y = \pm 1$	7.6911 %	0.2773 %
$z = \pm 1$	7.6911 %	0.2578 %
All faces	2.1450 %	0.0519 %
Average residual $(\bar{r})$	0.2366E-01	0.7605E-02
Maximum residual r_{max}	0.5197E-01	0.3068E-01

Table 10.1: $\bar{\epsilon}(\partial u / \partial n)$ (see (10.16)) and residuals $\bar{r}$, r_{max} for the initial cell configuration (Figure 10.4(a)) and the configuration obtained at the end of the *first* step of the adaptivity process using $\gamma = 1.0$ (Figure 10.4(b)) (from [28])

10.3.3   Initial Cell Configuration # 2 (96 Surface Cells)

The initial configuration # 2 is for the same physical cube with 16 uniform cells on each face (Figure 10.7(a)) with, as always, one node at the centroid of each cell. As before, the boundary value problem is solved using the BNM (equation (9.1)), and the results obtained are used in the HBNM (equation (9.12)). Figure 10.6 shows a comparison between the hypersingular residual (from equations (10.3) and (10.8)) and the exact local error ϵ in $u_{,j}$ (from equations (10.9) and (10.10)), computed for the initial configuration # 2 (Figure 10.7(a)). It can be clearly seen that the hypersingular residual tracks the exact error very accurately. In fact, the results for the finer cell configuration #2 are very similar to those shown in Figure 10.5 for the coarser initial cell configuration # 1.

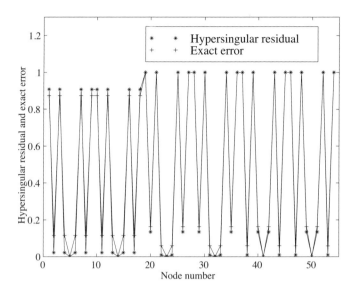

Figure 10.5: Comparison of hypersingular residual and exact local error ϵ in $u_{,j}$ for the initial configuration # 1 (54 cells, one node per cell). These quantities have been normalized by their respective maximum values, where $r_{max} = 0.5197 \times 10^{-1}$ and $\epsilon_{max} = 0.2051$ (from ([28])

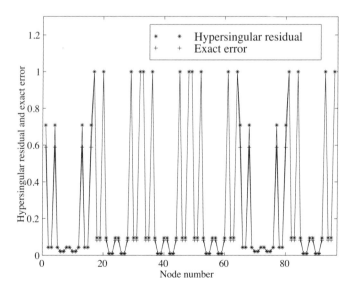

Figure 10.6: Comparison between hypersingular residual and exact local error ϵ in $u_{,j}$ for the initial configuration # 2 (96 cells, one node per cell). These quantities have been normalized by their respective maximum values, where $r_{max} = 0.1829 \times 10^{-1}$ and $\epsilon_{max} = 0.6223 \times 10^{-1}$ (from ([28])

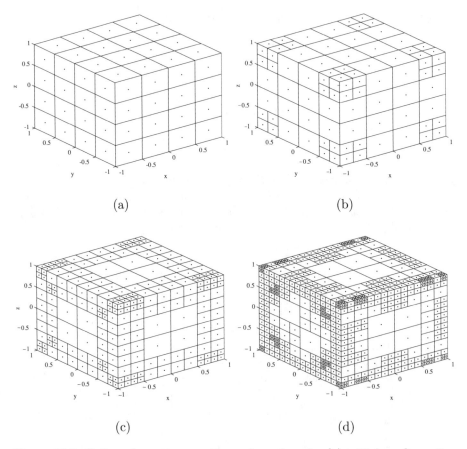

Figure 10.7: Cell configurations on the surface of a cube (a) initial configuration # 2 (96 surface cells), (b) first adapted step (168 cells), (c) second step (456 cells), (d) third step (1164 cells) (from ([28])

Adaptivity results. In order to obtain a better understanding of the adaptivity procedure, the local error e in $\partial u/\partial n$ (from equation (10.13)) is calculated on each of the faces of the cube. The iterative cell design cycle of Figure 10.2 is repeated three times using $\gamma = 0.5$ in equation (10.15) and starting from the initial configuration # 2 given in Figure 10.7(a). The resulting refined cell configurations are shown in Figures 10.7(b), (c) and (d), respectively. It is noted that the cell refinement should begin at the corners of the cube where the error in $\partial u/\partial n$ is the largest.

Figure 10.8 shows contour plots of the exact local error e in $\partial u/\partial n$ on the $y = -1$ face of the cube. The underlying cell structure on the face is also shown in the color plots. The resolution of these and subsequent contour plots is much finer than the corresponding cell discretization because the error is actually evaluated at a large number of points on the boundary (panels) of the body. These results confirm the observation made at the end of the previous paragraph regarding regions of large errors which demand a finer discretization. Thus, refinement occurs close to the edges and corners where the error in $\partial u/\partial n$ is largest.

Other relevant comments are in order. For the first step of the adaptive procedure (see Figure 10.7(b)), selected results are shown in Figure 10.9, which provides a comparison between the hypersingular residual (from equations (10.3) and (10.8)) and the exact local error ϵ (from equations (10.9) and (10.10)). The results are shown on the $x = -1$ and $z = 1$ faces as a representative sample of the results over the 168 nodes. It can be seen from Figure 10.9 that the hypersingular residual tracks the exact error reasonably well.

Figure 10.8(b) shows a contour plot for the exact local error e in $\partial u/\partial n$ on the $y = -1$ face of the cube for adapted cell configuration of Figure 10.7(b). Note that, due to the refinement procedure, the error in $\partial u/\partial n$ has reduced substantially, especially at the corners (cf Figures 10.8(a) and (b)).

Figures 10.8(c) and (d) show the exact local error e in $\partial u/\partial n$ on the $y = -1$ face of the cube for the adapted cell configurations consisting of 456 cells and 1164 cells, respectively (see Figures 10.9 (c) and (d)). Comparing the contour plots of Figures 10.8(a)-(d), one can readily verify that the error in $\partial u/\partial n$ decreases substantially during the adaptive process. It is interesting to note that the absolute value of the exact solution (equation (10.18)) has the same functional dependence on the $y = -1$ and $z = 1$ faces and different on the $x = -1$ face of the cube. Step 1 (Figure 10.7(b)) is not sensitive to this fact, however, Steps 2 (Figure 10.7(c)) and 3 (Figure 10.7(d)) of the adaptive procedure are. This is a tribute to the quality of residuals as error estimates.

Table 10.2 summarizes the results of the adaptive process for the cube problem starting with the initial cell configuration consisting of 96 cells (Figure 10.7(a)). Note that, on the faces $x = \pm 1$, $\bar{\epsilon}(\partial u/\partial n)$ (from equation (10.16)) increases from the initial configuration to Step 1, and from Step 1 to Step 2. However, $\bar{\epsilon}(\partial u/\partial n)$ finally decreases from Step 2 to Step 3 and reaches its lowest value at this step, which has the sophisticated cell pattern of Figure 10.7(d). On the faces $y = \pm 1$ and $z = \pm 1$, $\bar{\epsilon}(\partial u/\partial n)$ monotonically decreases as the number

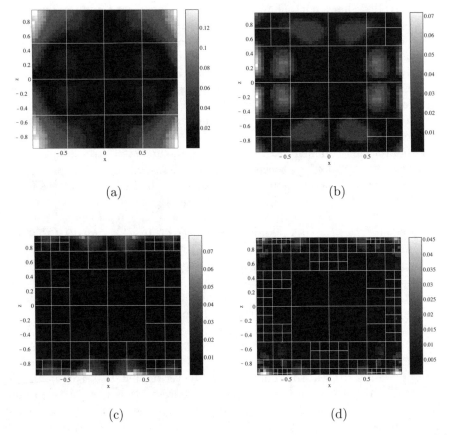

Figure 10.8: Error in $\partial u/\partial n$ ($e(\partial u/\partial n)$) on the face $y = -1$ of the cube (a) initial configuration # 2 (96 surface cells), (b) first adapted step (168 cells), (c) second step (456 cells), (d) third step (1164 cells) (from ([28])

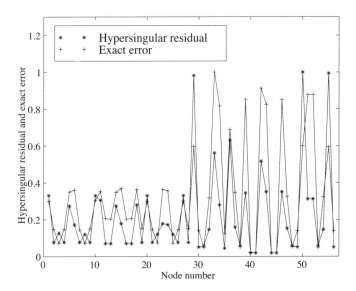

Figure 10.9: Comparison of hypersingular residual and exact local error ϵ in $u_{,j}$ on the faces $x = -1$ and $z = 1$ of the cube of Figure 10.7(b) (*first* step of the adaptive procedure). The quantities have been normalized by their respective maximum values, where $r_{max} = 0.1567 \times 10^{-1}$ and $\epsilon_{max} = 0.3645 \times 10^{-1}$ (from ([28])

Output parameters	Initial	Step 1	Step 2	Step 3
Number of cells	96	168	456	1164
$x = \pm 1$	0.0759 %	0.1062 %	0.1135 %	0.0438 %
$y = \pm 1$	1.0654 %	0.2785 %	0.2089 %	0.0551 %
$z = \pm 1$	1.0696 %	0.2781 %	0.2091 %	0.0551 %
All faces	0.1899 %	0.1269 %	0.1247 %	0.0451 %
Average residual $\bar{r}$	0.4963E-02	0.3661E-02	0.5643E-03	0.1811E-03
Max. residual r_{max}	0.1829E-01	0.1567E-01	0.3579E-02	0.2537E-02

Table 10.2: L_2 error in $\partial u/\partial n$ ($\bar{\epsilon}(\partial u/\partial n)$) and residuals $\bar{r}$, r_{max} for the various steps of the adaptivity process starting with the initial cell configuration consisting of 96 cells with one node per cell (Figure 10.7(a)). Here $\gamma = 0.5$ is used for the cell refinement of the cube (from [28])

of adaptive cycles increases. Moreover, as expected, the global $\bar{\epsilon}(\partial u/\partial n)$ for "all faces," as well as the average and maximum residuals, decrease as the adaptive process progresses.

10.4   One-Step Adaptive Cell Refinement

The previous section has dealt with an iterative adaptive technique for cell refinement (h−version). Here the interest is on developing a simple ONE-step algorithm for cell refinement in the meshless BNM setting. The flowchart of Figure 10.10 illustrates this idea which is based on the concept of refinement level (RL) employed by Krishnamoorthy and Umesh [74].

Refinement strategy.   Figures 10.11(a) and (b) show that different degrees of refinement are carried out for different values of the refinement level. From these figures, the expression relating the final cell size h_f to the refinement level RL is:

$$h_f = \frac{h_i}{2^{RL}} \tag{10.19}$$

where h_i denotes the initial cell size. *Assuming* that the rate of convergence of the error is $\mathcal{O}(h^p)$, where h is a characteristic cell size in the area covered by the cells, which are of order p, and setting the error estimate equal to $\eta = r/(\gamma \bar{r})$ (see equations (10.14) and (10.15)), one obtains:

$$h_f = \frac{h_i}{\eta^{1/p}} \tag{10.20}$$

From equations (10.19) and (10.20), the RL is given by:

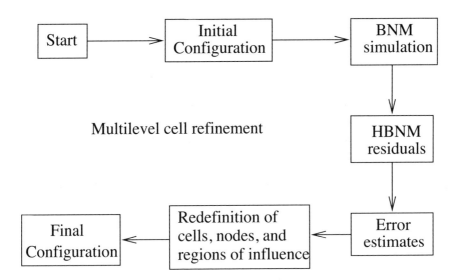

Figure 10.10: ONE-step adaptive BEM algorithm based on multilevel cell refinement (from ([28])

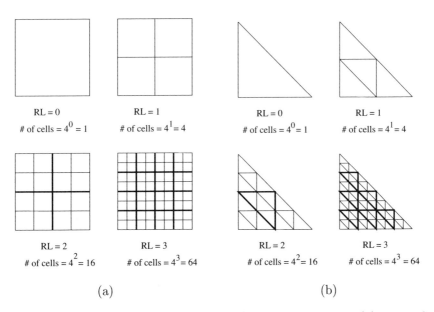

Figure 10.11: Refinement level RL using (a) rectangular and (b) triangular cells. The bold lines illustrate the idea of cell structure embedding (from ([28])

$$RL = \begin{cases} \frac{\log \eta}{p \log 2} & \text{for } \eta \geq 1 \\ 0 & \text{for } \eta < 1 \end{cases} \tag{10.21}$$

where p is order of the interpolating function. For the interpolation procedure used in this work, $p = m$. The second condition in equation (10.21) is enforced because cell structure coarsening is not considered in this work.

This idea of ONE-step refinement is applied below to the cube problem of Section 10.3. Errors are again estimated by means of hypersingular residuals.

10.4.1   Initial Cell Configuration # 1 (54 Surface Cells)

The ONE-step multilevel strategy is implemented on the cube of Figure 10.12(a), which consists of 54 cells with one node per cell. The boundary value problem is solved using the BNM (equation (9.1)) by imposing the exact solution in equation (10.18) as Dirichlet boundary conditions. The hypersingular residual (from equations (9.12 and (10.8)) is obtained and then the multilevel refinement procedure is carried out using $\gamma = 0.15$. The cell structure obtained in ONE-step is shown in Figure 10.12(b), which consists of 438 cells with one node per cell. Table 10.3 shows a comparison of the results from the ONE-step multilevel refinement scheme starting with the configuration of Figure 10.12(a) and ending with the configuration of Figure 10.12(b). This table shows that $\bar{\epsilon}(\partial u/\partial n)$ and the residual consistently decrease with refinement.

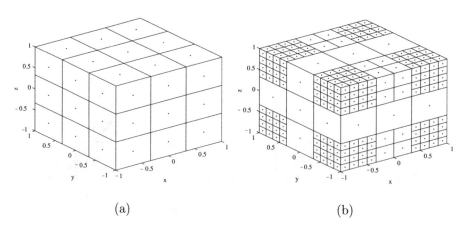

(a)   (b)

Figure 10.12: ONE-step multilevel cell refinement for the cube problem. (a) Initial configuration # 1 with 54 cells. (b) Adapted configuration with 438 cells using $\gamma = 0.15$ (from ([28])

Output parameters	Initial	Final
Number of cells	54	438
$x = \pm 1$	1.4209 %	0.0411 %
$y = \pm 1$	7.6911 %	0.0339 %
$z = \pm 1$	7.6911 %	0.0343 %
All faces	2.1450 %	0.0403 %
Average residual $\bar{r}$	0.2366E-01	0.4615E-03
Maximum residual r_{max}	0.5197E-01	0.2618E-01

Table 10.3: $\bar{\epsilon}(\partial u/\partial n)$ and residuals $\bar{r}$, r_{max} for the initial configuration (Figure 10.12(a)), and the final configuration (Figure 10.12(b)) obtained by the multilevel refinement strategy with $\gamma = 0.15$ (from [28])

10.4.2 Initial Cell Configuration # 2 (96 Surface Cells)

The BNM (equation (9.1)) is used to solve the boundary value problem using equation (10.18) as the exact solution, and the hypersingular residual is obtained by means of equations (9.12) and (10.8). Multilevel refinement is carried out using $\gamma = 0.15$. The cell structure obtained in ONE-step is shown in Figure 10.13(b), which consists of 1764 cells with one node per cell.

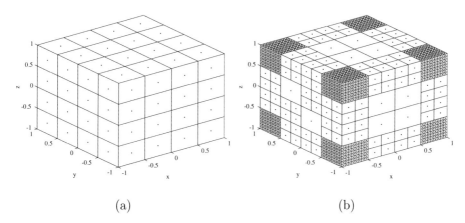

(a)   (b)

Figure 10.13: ONE-step multilevel cell refinement for the cube problem (a) initial configuration # 2 with 96 cells (b) adapted configuration with 1764 cells using $\gamma = 0.15$ (from ([28])

Table 10.4 summarizes the results for $\bar{\epsilon}(\partial u/\partial n)$ and the residual for the multilevel refinement strategy starting from the initial configuration # 2 of 96 cells (see Figure 10.13(a)). Qualitatively, these results are analogous to those of Table 10.2 for the progressive adaptive refinement, which includes the peculiarity

Output parameters	Initial	Final
Number of cells	96	1764
$x = \pm 1$	0.0759 %	0.1034 %
$y = \pm 1$	1.0654 %	0.2400 %
$z = \pm 1$	1.0696 %	0.2400 %
All faces	0.1899 %	0.1169 %
Average residual $(\bar{r})$	0.4963E-02	0.3370E-03
Maximum residual (r_{max})	0.1829E-01	0.1535E-01

Table 10.4: $\bar{\epsilon}(\partial u/\partial n)$ and residual for the initial configuration (Figure 10.13(a)), and the final configuration (Figure 10.13(b)) obtained by the multilevel refinement strategy with $\gamma = 0.15$ (from [28])

observed on the $x = \pm 1$ faces of the cube. Moreover, the remarks concerning Table 10.2 also hold for explaining the results of Table 10.4. Therefore, for further explanations, the reader is referred to Section 10.3.

Chapter 11

ADAPTIVITY FOR 3-D LINEAR ELASTICITY

The subject of this chapter is error analysis and adaptivity with the BNM, as applied to 3-D linear elasticity. Please see the introduction to Chapter 10.

11.1 Hypersingular and Singular Residuals

11.1.1 The Hypersingular Residual

Let the BNM (equation (9.24)) for elasticity be written in operator form as:

$$\mathcal{L}_{BNM}(u_k, \tau_k) = 0 \quad ; \quad k = 1, 2, 3 \tag{11.1}$$

with the *numerical solution* (u_k^*, τ_k^*). Also, the HBNM (equation (9.27)) is written in operator form as:

$$\mathcal{L}_{HBNM}(u_k, \tau_k) = 0 \quad ; \quad k = 1, 2, 3 \tag{11.2}$$

This time, the stress residual is defined from the stress HBNM (equation (9.27)) as,

$$r_{ij} \equiv residual(\sigma_{ij}) = \mathcal{L}_{HBNM}(u_k^*, \tau_k^*) \quad ; \quad k = 1, 2, 3 \tag{11.3}$$

This idea is illustrated in Figure 10.1(a).

It has been proved in [96] and [127] for the BIE that, under certain favorable conditions, real positive constants c_1 and c_2 exist such that:

$$c_1 r \leq \epsilon \leq c_2 r \tag{11.4}$$

where r is some scalar measure of a hypersingular residual and ϵ is a scalar measure of the exact local error. Thus, a hypersingular residual is expected to provide a good estimate of the local error on a boundary element. It should be mentioned here that the definitions of the residuals used in [96] and [127] are analogous to, but different in detail from, the ones proposed in this chapter.

11.1.2   The Singular Residual

From equation (9.28):

$$\mathcal{L}_{HBNM}(u_k^o, \tau_k^o) = 0 \quad ; \quad k = 1, 2, 3 \tag{11.5}$$

and from equation (9.24)

$$r_i \equiv residual(u_i) = \mathcal{L}_{BNM}(u_k^o, \tau_k^o) \quad ; \quad k = 1, 2, 3 \tag{11.6}$$

This idea is illustrated in Figure 10.1(b).

The above formulation for singular and hypersingular residuals is a generalization of the earlier work by Menon et al. [96] in the sense that Dirichlet, Neumann and mixed problems require separate prescriptions in [96], while the current work presents a unified residual formulation.

11.2   Error Estimation and Adaptive Strategy

The h−version iterative self-adaptive procedure employed in this work is presented in the flowchart - Figure 10.2. The goal is to efficiently develop a final cell configuration which leads to a reliable numerical solution, in as simple a manner as possible.

11.2.1   Local Residuals and Errors - Hypersingular Residual Approach

From equation (11.3) :

$$r_{ij} = residual(\sigma_{ij}) \tag{11.7}$$

A scalar residual measure is defined as:

$$r = r_{ij} r_{ij} \tag{11.8}$$

The exact local error in stress is defined as:

$$\epsilon_{ij} = \sigma_{ij}^{(exact)} - \sigma_{ij}^{(numerical)} \tag{11.9}$$

and the corresponding scalar measure is defined as:

$$\epsilon = \epsilon_{ij} \epsilon_{ij} \tag{11.10}$$

Equations (11.8) and (11.10) are used to compute the hypersingular residual and exact error, respectively, in the stress σ_{ij}, at each node, for problems in linear elasticity.

11.2.2 Local Residuals and Errors - Singular Residual Approach

From equation (11.6):

$$r_i = residual(u_i) \tag{11.11}$$

so that a scalar residual measure is:

$$r = r_i r_i \tag{11.12}$$

The exact local error in u_i is defined as:

$$\epsilon_i = u_i^{(exact)} - u_i^{(numerical)} \tag{11.13}$$

with a corresponding scalar measure:

$$\epsilon = \epsilon_i \epsilon_i \tag{11.14}$$

Equations (11.12) and (11.14) are used to obtain the singular residual and exact error, respectively, in the displacement u_i, at each node, for elasticity problems.

The scalar residual measures, defined above, evaluated at nodes, are used as error estimators. In all the adaptivity examples presented in this chapter, one node is used for each cell and is placed at its centroid. The scalar residual measure at this centroidal node is used as an error estimator for that cell. A comparison of the residual r and exact error ϵ demonstrates the effectiveness of residuals as error estimates.

11.2.3 Cell Refinement Global Error Estimation and Stopping Criterion

The algorithms used here are the same as those described in Sections 10.2.3 and 10.2.4 in Chapter 10.

11.3 Progressively Adaptive Solutions: Pulling a Rod

The problem under consideration here is the stretching of a cylindrical elastic rod clamped at one end (see Figure 11.1(a)). This time, the roles of the BNM and HBNM are reversed, i.e. the HBNM (9.28) is used to solve the boundary value problem while the singular residuals are obtained from the standard BNM (9.24) (see Figure 10.1(b)). This same problem has been addressed with the BCM in Section 7.3.1 of Chapter 7.

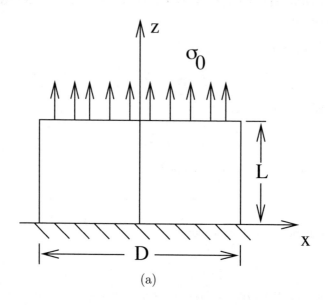

(a)

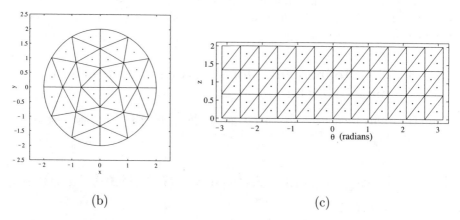

(b)   (c)

Figure 11.1: Stretching of a short clamped cylindrical rod by a uniform tensile load (a) physical situation : L = 2.0, D = 4.0, ν = 0.25, E = 1.0, σ_0 = 1.0 (b) and (c) initial cell configuration with 144 cells (one node per cell) - (b) clamped and loaded faces (c) curved surface of the cylindrical rod mapped onto the (z, θ) plane (from [28])

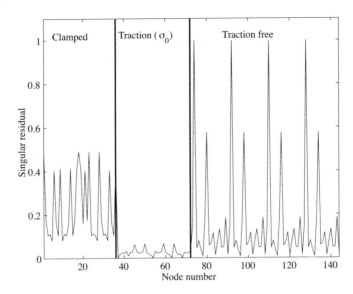

Figure 11.2: Singular residual for the initial configuration of 144 cells on the cylindrical rod of Figure 11.1 (b) and (c). The residual has been normalized with respect to its maximum value, $r_{max} = 0.2419 \times 10^{-2}$ (from [28])

11.3.1 Initial Cell Configuration

The geometric and material parameters chosen are: $E = 1.0$, $\nu = 0.25$, $\sigma_0 = 1.0$, $L = 2.0$, and $D = 4.0$. Figures 11.1(b) and (c) show the initial cell configuration on the clamped and loaded faces and on the curved surface of the rod. The boundary value problem is solved by prescribing tractions on the top face of the cylinder, with the bottom surface completely clamped, and the curved surface traction free. Upon obtaining the solution to the boundary value problem, the singular residual is obtained at each node. Figure 11.2 shows the singular residual (from equations (9.28), (11.6), and (11.12)) obtained for the initial cell configuration (144 cells). It can be seen that the residual is considerably higher on the clamped face and on the curved surface near the clamped face, than on the loaded face. This is to be expected considering the physical nature of the problem at hand which has a singularity on the bounding circle of the clamped face [37, 137].

11.3.2 Adaptivity Results

The adaptive strategy is carried out according to the flow chart of Figure 10.2. However, the boxes for the "BNM simulation" and "HBNM residuals" are replaced by "HBNM simulation" and "BNM residuals," according to the scheme of Figure 10.1(b). Since the singular residual is higher on the clamped and curved faces, most of the subdivision of cells occurs on those faces. The curved surface of the cylinder near the clamped surface is refined due to the singularity at the edge of the clamped face. However, the top face (the loaded face) is NOT refined at all and so the cell structure on that face remains as shown in Figure 11.1(b).

Three steps of adaptivity are pursued using $\gamma = 1.25$ in equation (10.15) and starting from the initial configuration of Figures 11.1(b) and (c). The resulting refined cell configurations are shown in Figures 11.3 - 11.5. As expected, Figures 11.3(a), 11.4(a) and 11.5(a) show that the loaded face is not refined at all and remains as in the initial configuration (Figure 11.1(b)). On the clamped face, a comparison of Figures 11.1(b), 11.3(b), 11.4(b), and 11.5(b) indicates that cell refinement only takes place near the edge of the face, which is the region where gradients in stresses are largest.

Figures 11.1(c), 11.3(c), 11.4(c), and 11.5(c) show the progressive refinement on the curved surface of the cylinder. One can observe that refinement primarily occurs along the curved surface near the clamped edge of the cylindrical rod. Note that, for the initial configuration (Figure 11.1 (b) and (c)), the number of subdivisions along the edge of the top and bottom faces is the same as the number of subdivisions along the edge of the curved surface (12 subdivisions). However, when adaptivity is carried out, a significant mismatch in the number of subdivisions is created at every adaptive step. This does not present any problem for the meshless method, and such freedom in modeling is expected to be especially advantageous in analyzing problems with complicated geometry.

11.4   One-Step Adaptive Cell Refinement

The ONE-step procedure outlined in Section 10.4 is demonstrated here for the same problem as above (Section 11.3). The initial configuration is the same as that of Figure 11.1(b) and (c). The present study also employs singular residuals (equation (9.24)) for error estimation (see Section 11.3) rather than hypersingular residuals (equation (9.27)). This is similar to the study carried out for the same problem by means of iterative adaptive cell refinement. The results for the multilevel refinement for the original problem of Figure 11.1(a) and (b) are given in Figure 11.6. Comparing these results with the ones of Section 11.3 (Figures 11.3-11.5), one verifies that the overall trends are quite similar in both situations. However, two main differences are noticeable. First, the loaded face is refined (a little) here (see Figures 11.1(b) and 11.6(a)), while it is not refined at all in the adapted configurations shown in Figures 11.3(a), 11.4(a) and 11.5(a). Second, as expected, the multilevel cell refinement does not allow the smooth cell gradation which occurs in progressively adapted cell configurations (see Figures 11.5 and 11.6). Nevertheless, such gradation, which is essential for mesh-based methods, is not required at all in the present meshless methods (BNM and HBNM).

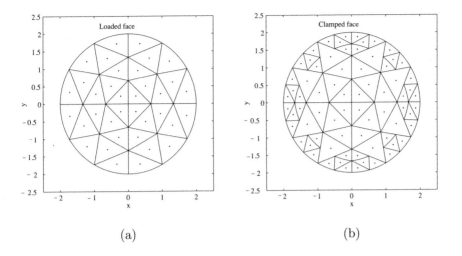

(a)   (b)

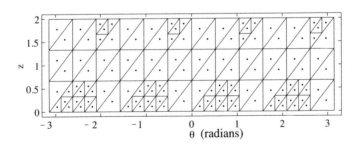

(c)

Figure 11.3: Short clamped cylindrical rod Step 1 : Adapted configuration consisting of 228 cells obtained with $\gamma = 1.25$; (a) loaded face (b) clamped face (c) curved surface of the rod (from [28])

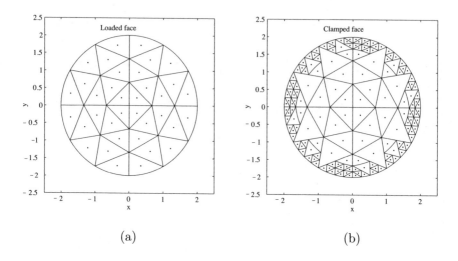

(a)

(b)

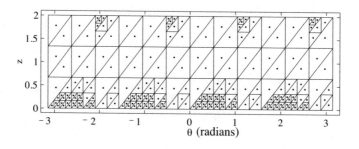

(c)

Figure 11.4: Short clamped cylindrical rod Step 2 : Adapted configuration consisting of 324 cells obtained with $\gamma = 1.25$; (a) loaded face (b) clamped face (c) curved surface of the rod (from [28])

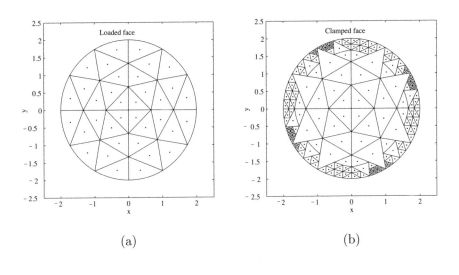

(a)   (b)

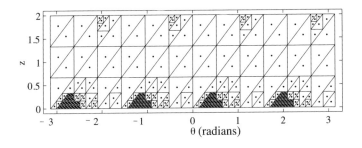

(c)

Figure 11.5: Short clamped cylindrical rod Step 3 : Adapted configuration consisting of 576 cells obtained with $\gamma = 1.25$; (a) loaded face (b) clamped face (c) curved surface of rod (from [28])

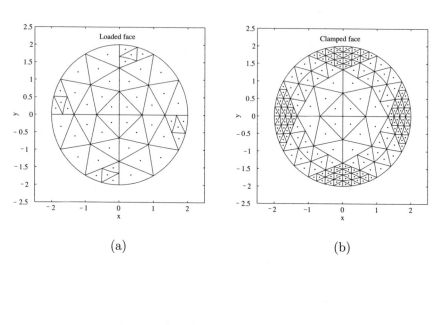

(a)   (b)

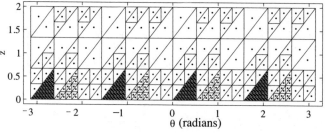

(c)

Figure 11.6: ONE-step multilevel cell refinement for short clamped cylinder: (a) loaded face (b) clamped face (c) curved surface of cylinder. The initial cell configuration is shown in Figure 11.1(b) and (c) (from [28])

Bibliography

[1] Aluru, N.R. and White, J. (1997). An efficient numerical technique for electromechanical simulation of complicated microelectromechanical structures. *Sensors and Actuators A* 58:1-11.

[2] Aluru, N.R. and Li, G. (2001). Finite cloud method: a true meshless technique based on a fixed reproducing kernel approximation. *International Journal for Numerical Methods in Engineering* 50:2373-2410.

[3] Atluri, S.N. and Shen, S. (2002). *The Meshless Local Petrov-Galerkin (MLPG) Method.* Tech Science Press, Encino, CA.

[4] Banerjee, P.K. (1994). *The Boundary Element Methods in Engineering.* McGraw Hill Europe, Maidenhead, Berkshire, UK.

[5] Banichuk, N.V. (1983). *Problems and Methods of Optimal Structural Design.* Plenum, New York.

[6] Bao, Z. and Mukherjee, S. (2004). Electrostatic BEM for MEMS with thin conducting plates and shells. *Engineering Analysis with Boundary Elements.* 28:1427-1435.

[7] Bao, Z. and Mukherjee, S. (2005). Electrostatic BEM for MEMS with thin beams. *Communications in Numerical Methods in Engineering.* In press.

[8] Bao, Z., Mukherjee, S., Roman, M. and Aubry, N. (2004). Nonlinear vibrations of beams strings plates and membranes without initial tension. *ASME Journal of Applied Mechanics* 71:551-559.

[9] Becker, A.A. (1992). *The Boundary Element Method in Engineering.* McGraw Hill International, Singapore.

[10] Belytschko, T., Lu, Y.Y. and Gu, L. (1994). Element-free Galerkin methods. *International Journal for Numerical Methods in Engineering* 37:229-256.

[11] Belytschko, T., Krongauz, Y., Organ, D., Fleming, M. and Krysl, P. (1996). Meshless methods: An overview and recent developments. *Computer Methods in Applied Mechanics and Engineering* 139:3-47.

[12] Bobaru, F. and Mukherjee, S. (2001). Shape sensitivity analysis and shape optimization in planar elasticity using the element-free Galerkin method. *Computer Methods in Applied Mechanics and Engineering* 190:4319-4337.

[13] Bobaru, F. and Mukherjee, S. (2002). Meshless approach to shape optimization of linear thermoelastic solids. *International Journal for Numerical Methods in Engineering* 53:765-796.

[14] Bonnet, M. (1995). *Boundary Integral Equation Methods for Solids and Fluids.* Wiley, Chichester, UK.

[15] Bonnet, M. and Xiao, H. (1995). Computation of energy release rate using material differentiation of elastic BIE for 3-D elastic fracture. *Engineering Analysis with Boundary Elements* 15:137-149.

[16] Brebbia, C.A. and Dominguez, J. (1992). *Boundary Elements: An Introductory Course.* 2nd edition. Computational Mechanics Publications, Southampton, UK, and McGraw Hill, New York.

[17] Bui, H.D. (1975). Application des potentiels élastiques à l'étude des fissures planes de forme arbitrare en milieu tridimensionnel. *Comptes Rendus de l'Académie des Sciences Paris Série A* 280:1157-1160.

[18] Carstensen, C. (1995). Adaptive boundary element methods and adaptive finite element methods and boundary element coupling. *Boundary Value Problems and Integral Equations in Nonsmooth Domains.* M. Costabel, M. Dauge and S. Nicaise eds., 47-58. Marcel Dekker.

[19] Carstensen, C. and Stephan, E.P. (1995). A posteriori error estimates for boundary element methods. *Mathematics of Computation* 64:483-500.

[20] Carstensen, C., Estep, D. and Stephan, E.P. (1995). h-adaptive boundary element schemes. *Computational Mechanics* 15:372-383.

[21] Carstensen, C. (1996). Efficiency of a posteriori BEM error estimates for first kind integral equations on quasi-uniform meshes. *Mathematics of Computation* 65:69-84.

[22] Chandra, A. and Mukherjee, S. (1997). *Boundary Element Methods in Manufacturing.* Oxford University Press, NY.

[23] Chati, M.K. (1999). *Meshless Standard and Hypersingular Boundary Node Method - Applications in Three-Dimensional Potential Theory and Linear Elasticity.* Ph.D. Dissertation, Cornell University, Ithaca, NY.

[24] Chati, M.K. and Mukherjee, S. (1999). Evaluation of gradients on the boundary using fully regularized hypersingular boundary integral equations. *Acta Mechanica* 135:41-55.

[25] Chati, M.K., Mukherjee, S. and Mukherjee, Y.X. (1999). The boundary node method for three-dimensional linear elasticity. *International Journal for Numerical Methods in Engineering* 46:1163-1184.

[26] Chati, M.K. and Mukherjee, S. (2000). The boundary node method for three-dimensional problems in potential theory. *International Journal for Numerical Methods in Engineering* 47:1523-1547.

[27] Chati, M.K., Mukherjee, S. and Paulino, G.H. (2001). The meshless hypersingular boundary node method for three-dimensional potential theory and linear elasticity problems. *Engineering Analysis with Boundary Elements* 25:639-653.

[28] Chati, M.K., Mukherjee, S. and Paulino, G.H. (2001). The meshless standard and hypersingular boundary node methods - applications to error estimation and adaptivity in three-dimensional problems. *International Journal for Numerical Methods in Engineering* 50:2233-2269.

[29] Chen, G. and Zhou, J. (1992). *Boundary Element Methods*. Academic Press.

[30] Chen, J.T. and Hong, H.K. (1999). Review of dual boundary element methods with emphasis on hypersingular integrals and divergent series. *ASME Applied Mechanics Reviews* 52:17-33.

[31] Chen, S.Y., Tang, W.X., Zhou, S.J., Leng, X. and Zheng, W. (1999). Evaluation of J integrals and stress intensity factors in a 2D quadratic boundary contour method. *Communications in Numerical Methods in Engineering* 15:91-100.

[32] Chen, W. and Tanaka, M. (2002). A meshless, integration-free, boundary-only RBF technique. *Computers and Mathematics with Applications* 43:379-391.

[33] Choi, J.H. and Kwak, B.M. (1988). Boundary integral equation method for shape optimization of elastic structures. *International Journal for Numerical Methods in Engineering* 26:1579-1595.

[34] Ciarlet, P.G. (1991). Basic error estimates for elliptic problems. *Finite Element Methods (Part 1) - Vol. 2 of Handbook of Numerical Analysis*. P. G. Ciarlet and J. L. Lions eds., Elsevier Science Publishers.

[35] Costabel, M. and Stephan, E. (1985). Boundary integral equations for mixed boundary value problems in polygonal domains and Galerkin approximation. *Mathematical Models and Methods in Mechanics Vol. 15*. W. Fiszdon and K. Wilmanski eds., Banach Center Publications.

[36] Costabel, M., Dauge, M. and Nicaise, S. eds. (1995). *Boundary Value Problems and Integral Equations in Non-smooth Domains. Number 167 in Lecture Notes in Pure and Applied Mathematics*. Marcel Dekker Publications.

[37] Cruse, T.A. (1969). Numerical solutions in three-dimensional elastostatics. *International Journal of Solids and Structures* 5:1259-1274.

[38] Cruse, T.A. (1988). *Boundary Element Analysis in Computational Fracture Mechanics.* Kluwer, Dordrecht, The Netherlands.

[39] Cruse, T.A. and Richardson, J.D. (1996). Non-singular Somigliana stress identities in elasticity. *International Journal for Numerical Methods in Engineering* 39:3273-3304.

[40] De, S. and Bathe, K.J. (2000). The method of finite spheres. *Computational Mechanics* 25:329-345.

[41] Dolbow, J., Möes, N. and Belytschko, T. (2000). Modeling fracture in Mindlin-Reissner plates with the extended finite element method. *International Journal of Solids and Structures* 37:7161-7183.

[42] Duarte, C.A.M. and Oden, J.T. (1996). H-p clouds - an h-p meshless method. *Numerical Methods in Partial Differential Equations* 12:673-705.

[43] Duarte, C.A.M. and Oden, J.T. (1996). An h-p adaptive method using clouds. *Computer Methods in Applied Mechanics and Engineering* 139:237-262.

[44] Eriksson, K., Estep, D., Hansbo, P. and Johnson, C. (1995). Introduction to adaptive methods for partial differential equations. *Acta Numerica* Chapter 3. A. Iserles ed., 105-158, Cambridge University Press, UK.

[45] Feistaur, M., Hsiao, G.C. and Kleinman, R.E. (1996). Asymptotic and a posteriori error estimates for boundary element solutions of hypersingular integral equations. *SIAM Journal of Numerical Analysis* 32:666-685.

[46] Fung, Y-C. (1965). *Foundations of Solid Mechanics.* Prentice Hall, Englewood Cliffs, NJ.

[47] Gaul, L., Kögl, M. and Wagner, M. (2003). *Boundary Element Methods for Engineers and Scientists.* Springer Verlag, Heidelberg, Germany.

[48] Ghosh, N. and Mukherjee, S. (1987). A new boundary element method formulation for three dimensional problems in linear elasticity. *Acta Mechanica* 67:107-119.

[49] Gilbert, J.R., Legtenberg, R. and Senturia, S.D. (1995). 3D coupled electromechanics for MEMS - applications of CoSolve EM. *Proceedings of IEEE MEMS* pp.122-127.

[50] Glushkov, E., Glushkova, N. and Lapina, O. (1999). 3-D elastic stress analysis at polyhedral corner points. *International Journal of Solids and Structures* 36:1105-1128.

[51] Goldberg, M.A. and Bowman, H. (1998). Superconvergence and the use of the residual as an error estimator in the BEM - I. Theoretical Development. *Boundary Element Communications* 8:230-238.

[52] Gowrishankar, R. and Mukherjee, S. (2002). A "pure" boundary node method in potential theory. *Communications in Numerical Methods in Engineering* 18:411-427.

[53] Gray, L.J. (1989). Boundary element method for regions with thin internal cavities. *Engineering Analysis with Boundary Elements* 6:180-184.

[54] Gray, L.J., Martha, L.F. and Ingraffea, A.R. (1990). Hypersingular integrals in boundary element fracture analysis. *International Journal for Numerical Methods in Engineering* 29:1135-1158.

[55] Gray, L.J., Balakrishna, C. and Kane, J.H. (1995). Symmetric Galerkin fracture analysis. *Engineering Analysis with Boundary Elements* 15:103-109.

[56] Gray, L.J. and Paulino, G.H. (1997). Symmetric Galerkin boundary integral formulation for interface and multizone problems. *International Journal for Numerical Methods in Engineering* 40:3085-3101.

[57] Gray, L.J. and Paulino, G.H. (1997). Symmetric Galerkin boundary integral fracture analysis for plane orthotropic elasticity. *Computational Mechanics* 20:26-33.

[58] Gray, L.J. and Paulino, G.H. (1998). Crack tip interpolation revisited. *SIAM Journal of Applied Mathematics* 58:428-455.

[59] Guidera, J.T. and Lardner, R.W. (1975). Penny shaped cracks. *Journal of Elasticity* 5:59-73.

[60] Guiggiani, M. (1994). Hypersingular formulation for boundary stress evaluation. *Engineering Analysis with Boundary Elements* 13:169-179.

[61] Harrington, R.F. (1993). *Field Computation by Moment Methods.* IEEE Press, Piscataway, NJ.

[62] Hartmann, F. (1989). *Introduction to Boundary Elements : Theory and Applications.* Springer Verlag, Berlin, New York.

[63] Haug, E.J., Choi, K.K. and Komkov, V. (1986). *Design Sensitivity Analysis of Structural Systems.* Academic Press, New York.

[64] Hayami, K. and Matsumoto, H. (1994). A numerical quadrature for nearly singular boundary element integrals. *Engineering Analysis with Boundary Elements* 13:143-154.

[65] Heck, A. (1993). *Introduction to Maple.* Springer Verlag, New York.

[66] Hughes, T.J.R. (2000). *The Finite Element Method - Linear Static and Dynamic Finite Element Analysis*. Dover, Mineola, NY.

[67] Kaljevic, I. and Saigal, S. (1997). An improved element free Galerkin formulation. *International Journal for Numerical Methods in Engineering* 40:2953-2974.

[68] Kane, J.H. (1994). *Boundary Element Analysis in Engineering Continuum Mechanics*. Prentice Hall, Englewood Cliffs, NJ.

[69] Kaplan, W. (1984). *Advanced Calculus* 3rd. ed. Addison-Wesley, Reading, MA.

[70] Kita, E. and Kamiya, N. (1994). Recent studies on adaptive boundary element methods. *Advances in Engineering Software* 19:21-32.

[71] Ko, S.C., Kim, Y.C., Lee, S.S., Choi, S.H. and Kim, S.R. (2003). Micromachined piezoelectric membrane acoustic device. *Sensors and Actuators A* 103:130-134.

[72] Kothnur, V., Mukherjee, S. and Mukherjee, Y.X. (1999). Two-dimensional linear elasticity by the boundary node method. *International Journal of Solids and Structures* 36:1129-1147.

[73] Kress, R. (1989). *Linear Integral Equations*. Springer Verlag.

[74] Krishnamoorthy, C.S. and Umesh, K.R. (1993). Adaptive mesh refinement for two-dimensional finite element stress analysis. *Computers and Structures* 48:121-133.

[75] Krishnasamy, G., Schmerr, L.W., Rudolphi, T.J. and Rizzo, F.J. (1990). Hypersingular boundary integral equations : some applications in acoustic and elastic wave scattering. *ASME Journal of Applied Mechanics* 57:404-414.

[76] Krishnasamy, G., Rizzo, F.J. and Rudolphi, T.J. (1992). Hypersingular boundary integral equations: their occurrence, interpretation, regularization and computation. *Developments in Boundary Element Methods-7*. P.K. Banerjee and S. Kobayashi eds., Elsevier Applied Science, London, pp. 207-252.

[77] Kulkarni, S.S., Telukunta, S. and Mukherjee, S. (2003). Application of an accelerated boundary-based mesh-free method to two-dimensional problems in potential theory. *Computational Mechanics* 32:240-249.

[78] Lancaster, P. and Salkauskas, K. (1990). *Curve and Surface Fitting - An Introduction*. Academic Press, London.

[79] Li, G. and Aluru, N.R. (2002). Boundary cloud method: a combined scattered point/boundary integral approach for boundary only analysis. *Computer Methods in Applied Mechanics and Engineering* 191:2337-2370.

[80] Li, G. and Aluru, N.R. (2003). A boundary cloud method with a cloud-by-cloud polynomial basis. *Engineering Analysis with Boundary Elements* 27:57-71.

[81] Liapis, S. (1995). A review of error estimation and adaptivity in the boundary element method. *Engineering Analysis with Boundary Elements* 14:315-323.

[82] Liu, G.R. (2002). *Mesh Free Methods - Moving beyond the Finite Element Method.* CRC Press, Boca Raton, FL.

[83] Liu, W.K., Chen, Y., Uras, R.A. and Chang, C.T. (1996). Generalized multiple scale reproducing kernel particle methods. *Computer Methods in Applied Mechanics and Engineering* 139:91-157.

[84] Liu, W.K., Chen, Y., Jun, S., Belytschko, T., Pan, C., Uras, R.A. and Chang, C.T. (1996). Overview and applications of the reproducing kernel particle methods. *Archives of Computational Methods in Engineering* 3, 3-80.

[85] Liu, Y.J. and Rizzo, F.J. (1993). Hypersingular boundary integral equations for radiation and scattering of elastic waves in three dimensions. *Computer Methods in Applied Mechanics and Engineering* 107:131-144.

[86] Liu, Y.J., Zhang, D. and Rizzo, F.J. (1993). Nearly singular and hypersingular integrals in the boundary element method. *Boundary Elements XV.* C.A. Brebbia and J.J. Rencis eds., 453-468, Computational Mechanics Publications, Southampton and Elsevier, Barking, Essex, UK.

[87] Liu, Y.J. (1998). Analysis of shell-like structures by the boundary element method based on 3-D elasticity : formulation and verification. *International Journal for Numerical Methods in Engineering* 41:541-558.

[88] Lutz, E.D. (1991). *Numerical Methods for Hypersingular and Near-Singular Boundary Integrals in Fracture Mechanics.* Ph.D. Dissertation, Cornell University, Ithaca, NY.

[89] Lutz, E.D., Ingraffea, A.R. and Gray, L.J. (1992). Use of 'simple solutions' for boundary integral methods in elasticity and fracture analysis. *International Journal for Numerical Methods in Engineering* 35:1737-1751.

[90] Lutz, E., Ye, W. and Mukherjee, S. (1998). Elimination of rigid body modes from discretized boundary integral equations. *International Journal of Solids and Structures* 35:4427-4436.

[91] Mackerele, J. (1993). Mesh generation and refinement for FEM and BEM - a bibliography (1990-1993). *Finite Elements in Analysis and Design* 15:177-188.

[92] Martin, P.A. and Rizzo, F.J. (1996). Hypersingular integrals : how smooth must the density be ? *International Journal for Numerical methods in Engineering* 39:687-704.

[93] Martin, P.A., Rizzo, F.J. and Cruse, T.A. (1998). Smoothness-relaxation strategies for singular and hypersingular integral equations. *International Journal for Numerical Methods in Engineering* 42:885-906.

[94] Mantič, V. (1993). A new formula for the C-matrix in the Somigliana identity. *Journal of Elasticity* 33:191-201.

[95] Menon, G. (1996). *Hypersingular Error Estimates in Boundary Element Methods*. M.S. Thesis, Cornell University, Ithaca, NY.

[96] Menon, G., Paulino, G.H. and Mukherjee, S. (1999). Analysis of hypersingular residual error estimates in boundary element methods for potential problems. *Computational Methods in Applied Mechanics and Engineering* 173:449-473.

[97] Moës, N., Dolbow, J. and Belytschko, T. (1999). A finite element method for crack growth without remeshing. *International Journal for Numerical Methods in Engineering* 46:131-150.

[98] Mukherjee, S. (1982). *Boundary Element Methods in Creep and Fracture*. Elsevier, London.

[99] Mukherjee, S. and Mukherjee, Y. X. (1998). The hypersingular boundary contour method for three-dimensional linear elasticity. *ASME Journal of Applied Mechanics* 65:300-309.

[100] Mukherjee, S., Shi, X. and Mukherjee, Y.X. (1999). Surface variables and their sensitivities in three-dimensional linear elasticity by the boundary contour method. *Computer Methods in Applied Mechanics and Engineering* 173:387-402.

[101] Mukherjee, S. (2000). CPV and HFP integrals and their applications in the boundary element method. *International Journal of Solids and Structures* 37:6623-6634.

[102] Mukherjee, S. (2000). Finite parts of singular and hypersingular integrals with irregular boundary source points. *Engineering Analysis with Boundary Elements* 24:767-776.

[103] Mukherjee, S., Shi, X. and Mukherjee, Y.X. (2000). Internal variables and their sensitivities in three-dimensional linear elasticity by the boundary

contour method. *Computer Methods in Applied Mechanics and Engineering* 187:289-306.

[104] Mukherjee, S., Chati, M.K. and Shi, X. (2000). Evaluation of nearly singular integrals in boundary element contour and node methods for three-dimensional linear elasticity. *International Journal of Solids and Structures* 37:7633-7654.

[105] Mukherjee, S. (2001). On boundary integral equations for cracked and for thin bodies. *Mathematics and Mechanics of Solids* 6:47-64.

[106] Mukherjee, S. (2002). Regularization of hypersingular boundary integral equations : a new approach for axisymmetric elasticity. *Engineering Analysis with Boundary Elements* 26:839-844.

[107] Mukherjee, Y.X. and Mukherjee, S. (1997). The boundary node method for potential problems. *International Journal for Numerical Methods in Engineering* 40:797-815.

[108] Mukherjee, Y.X. and Mukherjee, S. (1997). On boundary conditions in the element-free Galerkin method. *Computational Mechanics* 19:264-270.

[109] Mukherjee, Y.X., Mukherjee, S., Shi, X. and Nagarajan, A. (1997). The boundary contour method for three-dimensional linear elasticity with a new quadratic boundary element. *Engineering Analysis with Boundary Elements* 20:35-44.

[110] Mukherjee, Y.X., Shah, K. and Mukherjee, S. (1999). Thermoelastic fracture mechanics with regularized hypersingular boundary integral equations. *Engineering Analysis with Boundary Elements* 23:89-96.

[111] Mukherjee, Y.X. and Mukherjee, S. (2001). Error analysis and adaptivity in three-dimensional linear elasticity by the usual and hypersingular boundary contour method. *International Journal of Solids and Structures* 38:161-178.

[112] Nabors, K. and White, J. (1991). FastCap: a multi-pole accelerated 3-D capacitance extraction program. *IEEE Transactions on Computer-Aided Design of Integrated Circuits and Systems* 10:1447-1459.

[113] Nabors, K., Kim, S., White, J. and Senturia, S. (1992). *FastCap User's Guide*. MIT, Cambridge, MA.

[114] Nagarajan, A. and Mukherjee, S. (1993). A mapping method for numerical evaluation of two-dimensional integrals with $1/r$ singularity. *Computational Mechanics* 12:19-26.

[115] Nagarajan, A., Lutz, E.D. and Mukherjee, S. (1994). A novel boundary element method for linear elasticity with no numerical integration for 2-D

and line integrals for 3-D problems. *ASME Journal of Applied Mechanics* 61:264-269.

[116] Nagarajan, A., Mukherjee, S. and Lutz, E.D. (1996). The boundary contour method for three-dimensional linear elasticity. *ASME Journal of Applied Mechanics* 63:278-286.

[117] Nayroles, B., Touzot, G. and Villon, P. (1992). Generalizing the finite element method: diffuse approximation and diffuse elements. *Computational Mechanics* 10:307-318.

[118] Nishimura, N., Yoshida, K. and Kobayashi, S. (1999). A fast multipole boundary integral equation method for crack problems in 3D. *Engineering Analysis with Boundary Elements* 23:97-105.

[119] Novati, G. and Springhetti, R. (1999). A Galerkin boundary contour method for two-dimensional linear elasticity. *Computational Mechanics* 23:53-62.

[120] Oden, J.T., Duarte, C.A.M. and Zienkiewicz, O.C. (1998). A new cloud based *hp* finite element method. *Computer Methods in Applied Mechanics and Engineering* 153:117-126.

[121] París, F. and Cañas, J. (1997). *Boundary Element Method : Fundamentals and Applications.* Oxford University Press, UK.

[122] Paulino, G.H. (1995). *Novel Formulations of the Boundary Element Method for Fracture Mechanics and Error Estimation.* Ph.D. Dissertation, Cornell University, Ithaca, NY.

[123] Paulino, G.H., Gray, L.J. and Zarkian, V. (1996). Hypersingular residuals - a new approach for error estimation in the boundary element method. *International Journal for Numerical Methods in Engineering* 39:2005-2029.

[124] Paulino, G.H., Shi, F., Mukherjee, S. and Ramesh, P. (1997). Nodal sensitivities as error estimates in computational mechanics. *Acta Mechanica* 121:191-213.

[125] Paulino, G.H. and Gray, L.J. (1999). Galerkin residuals for adaptive symmetric-Galerkin boundary element methods. *ASCE Journal of Engineering Mechanics* 125:575-585.

[126] Paulino, G.H., Menezes, I.F.M., Cavalcante Neto, J.B. and Martha, L.F. (1999). A methodology for adaptive finite element analysis: towards an integrated computational environment. *Computational Mechanics* 23:361-388.

[127] Paulino, G.H., Menon, G. and Mukherjee S. (2001). Error estimation using hypersingular integrals in boundary element methods for linear elasticity. *Engineering Analysis with Boundary Elements* 25:523-534.

[128] Petryk, H. and Mróz, Z. (1986). Time derivatives of integrals and functionals defined on varying volume and surface elements. *Archives of Mechanics* 5-6:697-724.

[129] Phan, A.-V., Mukherjee, S. and Mayer, J.R.R. (1997). The boundary contour method for two-dimensional linear elasticity with quadratic boundary elements. *Computational Mechanics* 20:310-319.

[130] Phan, A.-V., Mukherjee, S. and Mayer, J.R.R. (1998). A boundary contour formulation for design sensitivity analysis in two-dimensional linear elasticity. *International Journal of Solids and Structures* 35:1981-1999.

[131] Phan, A.-V., Mukherjee, S. and Mayer, J.R.R. (1998). The hypersingular boundary contour method for two-dimensional linear elasticity. *Acta Mechanica* 130:209-225.

[132] Phan, A.-V., Mukherjee, S. and Mayer, J.R.R. (1998). Stresses, stress sensitivities and shape optimization in two-dimensional linear elasticity by the boundary contour method. *International Journal for Numerical Methods in Engineering* 42:1391-1407.

[133] Phan, A.-V. and Mukherjee, S. (1999). On design sensitivity analysis in linear elasticity by the boundary contour method. *Engineering Analysis with Boundary Elements* 23:195-199.

[134] Phan, A.-V. and Phan, T.-N. (1999). Structural shape optimization system using the two-dimensional boundary contour method. *Archive of Applied Mechanics* 69:481-489.

[135] Phan, A.-V. and Liu, Y.J. (2001). Boundary contour analysis of thin films and layered coatings. Sixth U.S. National Congress of Applied Mechanics, Detroit, MI.

[136] Phan, A.-V., Gray, L.J., Kaplan, T. and Phan, T.-N. (2002). The boundary contour method for two-dimensional Stokes flow and incompressible elastic materials. *Computational Mechanics* 28:425-433.

[137] Pickett, G. (1944). Application of the Fourier method to the solution of certain boundary value problems in the theory of elasticity. *Transactions of the ASME (Journal of Applied Mechanics)* 66:A176-A182.

[138] Poon, H., Mukherjee, S. and Ahmad, M.F. (1998). Use of simple solutions in regularizing hypersingular boundary integral equations in elastoplasticity. *ASME Journal of Applied Mechanics* 65:39-45.

[139] Radin, C. (1994). The pinwheel tilings of the plane. *Annals of Mathematics* 139:661-702.

[140] Rank, E. (1989). Adaptive h-, p-, and hp- versions for boundary integral element methods. *International Journal for Numerical Methods in Engineering* 28:1335-1349.

[141] Rizzo, F.J. (1967). An integral equation approach to boundary value problems of classical elastostatics. *Quarterly of Applied Mathematics* 25:83-95.

[142] Roman, M. and Aubry, N. (2003). Design and fabrication of electrically actuated synthetic microjets. *ASME Paper No. IMECE2003-41579*. American Society of Mechanical Engineers, New York.

[143] Rudolphi, T.J. (1991). The use of simple solutions in the regularization of hypersingular boundary integral equations. *Mathematical and Computer Modeling* 15:269-278.

[144] Saigal S. and Kane, J.H. (1990). Boundary-element shape optimization system for aircraft structural components. *AIAA Journal* 28:1203-1204.

[145] Sandgren, E. and Wu, S.J. (1988). Shape optimization using the boundary element method with substructuring. *International Journal for Numerical Methods in Engineering* 26:1913-1924.

[146] Schittkowski, K. (1986). NLPQL: A FORTRAN subroutine solving constrained nonlinear programming problems. *Annals of Operations Research* 5:485-500.

[147] Senturia, S.D., Harris, R.M., Johnson, B.P., Kim, S., Nabors, K., Shulman, M.A. and White, J.K. (1992). A computer-aided design system for microelectromechanical systems (MEMCAD). *Journal of Microelectromechanical Systems* 1:3-13.

[148] Shi, F., Ramesh, P. and Mukherjee, S. (1995). Simulation methods for micro-electro-mechanical structures (MEMS) with application to a microtweezer. *Computers and Structures* 56:769-783.

[149] Shi, F., Ramesh, P. and Mukherjee, S. (1996). Dynamic analysis of micro-electro-mechanical systems. *International Journal for Numerical Methods in Engineering* 39:4119-4139.

[150] Shi, X. and Mukherjee, S. (1999). Shape optimization in three-dimensional linear elasticity by the boundary contour method. *Engineering Analysis with Boundary Elements* 23:627-637.

[151] Sladek, J. and Sladek, V. (1986). Computation of stresses by BEM in 2-D elastostatics. *Acta Technica CSAV* 31:523-531.

[152] Sladek, J., Sladek, V. and Atluri, S.N. (2000). Local boundary integral equation (LBIE) method for solving problems of elasticity with nonhomogeneous material properties. *Computational Mechanics* 24:456-462.

[153] Sloan, I.H. (1976). Improvement by iteration for compact operator equations. *Mathematics of Computation* 30:758-764.

[154] Sloan, I.H. (1990). Superconvergence. *Numerical Solution of Integral Equations* Chapter 2. M. A. Goldberg ed., 35-70 Plenum Press, New York.

[155] Sloan, I.H. (1992). Error analysis in boundary integral methods. *Acta Numerica* Chapter 7. A. Iserles ed., 287-339 Cambridge University Press, UK.

[156] Stephan, E.P. (1996). The h-p boundary element method for solving 2- and 3- dimensional problems. *Computer Methods in Applied Mechanics and Engineering* 133:183-208.

[157] Strouboulis, T., Babuška, I. and Copps, K. (2000). The design and analysis of the generalized finite element method. *Computer Methods in Applied Mechanics and Engineering* 181:43-69.

[158] Sukumar, N., Moran, B. and Belytschko, T. (1998). The natural element method. *International Journal for Numerical Methods in Engineering* 43:839-887.

[159] Sukumar, N., Möes, N., Moran, B. and Belytschko, T. (2000). Extended finite element method for three-dimensional crack modeling. *International Journal for Numerical Methods in Engineering* 48:1549-1570.

[160] Sukumar, N., Moran, B., Semenov, A.Y. and Belytschko, T. (2001). Natural neighbour Galerkin methods. *International Journal for Numerical Methods in Engineering* 50:1-27.

[161] Tafreshi, A. and Fenner, R.T. (1995). General-purpose computer program for shape optimization of engineering structures using the boundary element method. *Computers and Structures* 56:713-720.

[162] Tanaka, M., Sladek, V. and Sladek, J. (1994). Regularization techniques applied to boundary element methods. *ASME Applied Mechanics Reviews* 47:457-499.

[163] Telukunta, S. and Mukherjee, S. (2004). An extended boundary node method for modeling normal derivative discontinuities in potential theory across corners and edges. *Engineering Analysis with Boundary Elements* 28:1099-1110.

[164] Telukunta, S. and Mukherjee, S. (2004). Extended boundary node method for three-dimensional problems in potential theory. *Computers and Structures*. Submitted.

[165] Telles, J.C.F. (1987). A self-adaptive co-ordinate transformation for efficient numerical evaluation of general boundary element integrals. *International Journal for Numerical Methods in Engineering* 24:959-973.

[166] Telles, J.C.F. and Oliveria, R.F. (1994). Third degree polynomial transformation for boundary element integrals : Further improvements. *Engineering Analysis with Boundary Elements* 13:135-141.

[167] Timoshenko, S.P. and Goodier, J.N. (1970). *Theory of Elasticity* 3rd. ed. McGraw Hill, New York.

[168] Toh, K.-C. and Mukherjee, S. (1994). Hypersingular and finite part integrals in the boundary element method. *International Journal of Solids and Structures* 31:2299-2312.

[169] Wei, X., Chandra, A., Leu, L.J. and Mukherjee, S. (1994). Shape optimization in elasticity and elasto-viscoplasticity by the boundary element method. *International Journal of Solids and Structures* 31:533-550.

[170] Wendland, W.L., Stephan, E.P. and Hsiao, G.C. (1979). On the integral equation method for the plane mixed boundary value problem for the Laplacian. *Mathematical Methods in the Applied Sciences* 1:265-321.

[171] Wendland, W.L. and Yu, D-H. (1988). Adaptive boundary element methods for strongly elliptic integral equations. *Numerische Mathematik* 53:539-558.

[172] Wendland, W.L. and Yu, D-H. (1992). A posteriori local error estimates of boundary element methods with some pseudo-differential equations on closed curves. *Journal of Computational Mathematics* 10:273-289.

[173] Wilde, A.J. and Aliabadi, M.H. (1998). Direct evaluation of boundary stresses in the 3D BEM of elastostatics. *Communications in Numerical Methods in Engineering* 14:505-517.

[174] Xu, Y. and Saigal, S. (1998). An element-free Galerkin formulation for stable crack growth in elastic solids. *Computer Methods in Applied Mechanics and Engineering* 154:331-343.

[175] Xu, Y. and Saigal, S. (1998). Element-free Galerkin study of steady quasi-static crack growth in plane strain tension in elastic-plastic materials. *Computational Mechanics* 21:276-282.

[176] Xu, Y. and Saigal, S. (1999). An element-free Galerkin analysis of steady dynamic growth of a mode I crack in elastic-plastic materials. *International Journal of Solids and Structures* 36:1045-1079.

[177] Yamazaki, K., Sakamoto, J. and Kitano, M. (1994). Three-dimensional shape optimization using the boundary element method. *AIAA Journal* 32:1295-1301.

[178] Yang, R.J. (1990). Component shape optimization using BEM. *Computers and Structures* 37:561-568.

[179] Yang, T.Y. (1986). *Finite Element Structural Analysis.* Prentice Hall, Englewood Cliffs, NJ.

[180] Yoshida, K., Nishimura, N. and Kobayashi, S. (2001). Application of a new fast multipole boundary integral equation method to crack problems in 3D. *Engineering Analysis with Boundary Elements* 25:239-247.

[181] Yu, D.-H. (1987). A posteriori error estimates and adaptive approaches for some boundary element methods. *Mathematical and Computational Aspects Vol. 1 of Boundary Elements IX.* C. Brebbia, W. L. Wendland and G. Kuhn eds., 241-256. Computational Mechanics Publications and Springer Verlag.

[182] Yu, D.-H. (1988). Self adaptive boundary element methods. *Zeitschrift für Angewandte Mathematik und Mechanik* 68:T435-T437.

[183] Zhang, Q. and Mukherjee, S. (1991). Design sensitivity coefficients for linear elastic bodies with zones and corners by the derivative boundary element method. *International Journal of Solids and Structures* 27:983-998.

[184] Zhao, Z. (1991). *Shape Design Sensitivity Analysis and Optimization Using the Boundary Element Method.* Springer, New York.

[185] Zhao, Z.Y. and Lan, S.R. (1999). Boundary stress calculation - a comparison study. *Computers and Structures* 71:77-85.

[186] Zhou, S.J., Sun, S.X. and Cao, Z.Y. (1998). The dual boundary contour method for two-dimensional crack problems. *International Journal of Fracture* 92:201-212.

[187] Zhou, S.J., Cao, Z.Y. and Sun, S.X. (1999). The traction boundary contour method for linear elasticity. *International Journal for Numerical Methods in Engineering* 46:1883-1895.

[188] Zhu T., Zhang, J-D. and Atluri, S.N. (1998). A local boundary integral equation (LBIE) method in computational mechanics, and a meshless discretization approach. *Computational Mechanics* 21:223-235.

[189] Zhu, T. (1999). A new meshless regular local boundary integral equation (MRLBIE) method. *International Journal for Numerical Methods in Engineering* 46:1237-1252.

[190] Zienkiewicz, O.C. and Taylor, R.L. (1994). *The Finite Element Method.* Vols. 1 and 2. McGraw Hill, Maidenhead, Berkshire, UK.

Index